THIS
EARTH OF
MANKIND

THIS EARTH OF MANKIND

Pramoedya Ananta Toer

Translated and with an Afterword by Max Lane

WILLIAM MORROW AND COMPANY, INC.

NEW YORK

First published in Indonesian by Hasta Mitra Publishing House, Jakarta, 1980, under the
   title *Bumi Manusia*.
Published by Penguin Books Australia, 1982, reprinted 1982
Revised 1991

Library of Congress Cataloging-in-Publication Data

Toer, Pramoedya Ananta, 1925–
    This earth of mankind / by Pramoedya Ananta Toer.
      p.   cm.

    1. Indonesia—History—1798–1942—Fiction.   I. Title.
PL5089.T8T46   1991
899'.22132—dc20   91-7346
                                                                        CIP

Printed in the United States of America

BOOK DESIGN BY NICOLA MAZZELLA

Translator's
Acknowledgment

Bumi Manusia was translated as *This Earth of Mankind* in 1981. Since then the translation has been revised twice, including for this U.S. edition. During the course of this process, which I am sure is still not finished, a large number of people have helped. Among those who have made contributions to refinement of the text are Kerry and Caroline Groves, the late R.F.X. Brissenden, Blanche del Alpuget, Jackie Yowell, and Elizabeth Flann. A special mention must be made of the late Dr. Geoff Blunden, who put considerable effort into editing the manuscript.

I must also express deep gratitude to Pramoedya Ananta Toer, Hasyim Rachman, and Yoesoef Isak, who together provided permission, support, and, most important of all, inspiration to finish this project. Indeed, I thank all my friends in Indonesia for the inspiration that they have been.

Finally, I thank Anna Nurfia and Melanie Purwitasari, who have been tolerant of my absences, either physical or mental, while I have been working on this project.

This narrow path has been trod many a time already, it's only that this time the journey is one to mark the way.

—P.A.T.

TRANSLATOR'S NOTE

This novel begins in Java in 1898. The port of Surabaya was an old town that grew up around a major harbor. On the ships frequenting the port were the famed products of the Spice Islands—the spices that originally attracted Europeans to the East Indies. There were also goods from all parts of Asia, machinery and products from nearly every country of Europe, as well as other natural products like rubber, coffee, sugar, and minerals, all heading back to Europe. They sailed southward, supplying the colonies soon to form Australia, while other ships returned, bringing, among other things, dairy cattle for Java's dairy industry.

The ships carried Dutch officials, businessmen, and adventurers from all over Europe, seeking their fortunes and perhaps ending up in the Dutch Colonial Army or prison. Arabs, Indians, and Chinese were always in port; many had long settled in Java. All these people brought with them aspects of the life of their own countries: their politics; their ideas on religion, philosophy, and morality; their prejudices and their hatreds; and sometimes their imagination and foresight.

Many of these ideas and ways of thinking came in other forms—books, for example. The narrator of our novel says, as he introduces his story, that of all the wonders of science, that at which he marveled most was printing, especially the printing of photographs. Books, newspapers, magazines, pamphlets were circulating everywhere; and soon there would be the international telegraph service.

This dynamic environment was no small problem for the government of the Dutch East Indies. That government presided over a colony the exploitation of whose resources made one of the smallest countries of Europe, Holland, one of its richest. This exploitation needed a special condition for its continuation: the maintenance of an attitude of acceptance on the part of the colonized and the governed. The colonizers' determination was that the native people, especially the toiling classes, of the Netherlands Indies should remain forever submerged in a culture of silence. This made exploitation easier and gave some Dutch their rationale for exhibiting the traditional colonial feelings of cultural arrogance and superiority.

The characters of this novel are set amid the tensions and contradictions created by the colliding of the liberating aspects of the expansion of capitalist industry and its technology on the one hand with the power of the colonial state on the other. In the center is the novel's narrator, Minke, an eighteen-year-old Native Javanese youth. He is the only Native in the Dutch High School (H.B.S.) in Surabaya. His attendance there points immediately to an outstanding success in primary school and, at least equally importantly, strong backing from an important Dutch or Native official, in Minke's case his grandparents.

His grandparents (though they never appear in the book, as they have passed away) symbolize the third element in the world that the book presents. This is the world of Java itself, or, more specifically, the world of those who for so long had dominated Java. Minke's mother, whom he refers to as Bunda (a soft, respectful, and loving term, inadequately translated in the text as Mother) and Minke's father, whom Minke knew only as a title, Father, are feudal Java's delegates in the novel. Through Minke's relations with his family, we are brought back to the ways of thinking and acting that the modern world and colonial powers faced when dealing with Java's feudal rulers. To the ways and

10

ideas of Minke's father and mother the forces of the imperialist economy pay no heed, while those of the colonial state seek only to manipulate them. But as his mother constantly reminds him, a Javanese must eventually come to terms with his own identity.

There are many more revelations for the reader about life in Java during this period. The reader may note, for example, the constant references to the languages used by the characters: Javanese, Dutch, Malay, Madurese, High Javanese. The use of language in this period was an important indicator of a person's social caste. Dutch, of course, was the language of the governing caste; Javanese, the Native language of the Javanese; and Madurese, the Native language of Madura (an island off Java). Malay was the language of interracial, or intercaste, communication (as many elite Javanese could speak Dutch), as well as the language of many Eurasians. Indeed, in situations in which the caste order needed to be emphasized, Natives were forbidden to use Dutch. Not only did colonialism install Dutch as the supreme caste language of Java, it helped reinforce and even exaggerate caste distinctions in the Native languages themselves, especially Javanese. The Javanese language already operated on at least three different levels, each based on the person to whom one was speaking or the person who was doing the speaking. This feudal stratification was given extra force as Javanese feudal notables, devoid of real political power in the face of the Dutch cannon and Dutch capital, channeled their oppressive energies into culture, something which Dutch cannon and capital were, in turn, frequently ready to buttress. The egalitarian and colloquial Javanese that was used in the palaces and royal houses of Java actually died out in this period. Only the masses of peasants and other toilers retained such an egalitarian Javanese.

The terms Native, Mixed-Blood, and Pure are capitalized because they do not simply identify racial origins, but manifest how, even in everyday life, race and caste dominated all of Netherlands Indies society. These categories were eventually given legal status. In the story that unfolds through this book, the true nonracial origins of this caste system—namely, the whim of colonial power and colonist—are exposed.

But this is not a book of revenge or hatred. Its spirit is not that of a simple denunciation. Pramoedya Ananta Toer set out to recreate the past through the telling of a story and the evocation of an atmosphere. He has actually brought to life in the first person the real ego of his main character, *a new historical personality in the process of being forged by history itself.* It is Minke, his psyche, his predicament, that is able to bring such a huge kaleidoscope of characters and stories together. The figure of Minke himself is announcing the coming of something that would envelop everybody, that would leave no part of society free of turmoil: a revolutionary future, the awakening of a people.

I have been guided by two principles in translating this novel. First, I have tried to cast it in a linguistic form that will facilitate the reader's enjoyment and easy reading of the novel, while remaining generally faithful to the author's text. Second, I have tried to avoid totally surrendering the translation of the text to the sovereignty that is sometimes given to the translator's language. At the same time the peculiar social and political history of language in Indonesia has meant that some linguistic phenomena have not been reproducible. This is especially so as regards the play between languages and levels of language.

A glossary of Javanese terms appears at the end of the book.

Max Lane
Jakarta and Canberra, 1981 and 1990

THIS
EARTH OF
MANKIND

1

People called me Minke.

My own name . . . for the time being I need not tell it. Not because I'm crazy for mystery. I've thought about it quite a lot: I don't yet really need to reveal who I am before the eyes of others.

In the beginning I wrote these short notes during a period of mourning: She had left me, who could tell if only for a while or forever? (At the time I didn't know how things would turn out.) That eternally harassing, tantalizing future. Mystery! We will all eventually arrive there—willing or unwilling, with all our soul and body. And too often it proves to be a great despot. And so, in the end, I arrived too. Whether the future is a kind or a cruel god is, of course, its own affair: Humanity too often claps with just one hand.

Thirteen years later I read and studied these short notes over again. I merged them together with dreams, imaginings. Naturally they became different from the original. Different? But that doesn't matter!

And here is how they turned out.

2

I was still very young, just the age of a corn plant, yet I had already experienced modern learning and science: They had bestowed upon me a blessing whose beauty was beyond description.

The director of my school once told my class: Your teachers have given you a very broad general knowledge, much broader than that received by students of the same level in many of the European countries.

Naturally this breast of mine swelled. I'd never been to Europe, so I did not know if the director was telling the truth or not. But because it pleased me, I decided to believe him. And, further, all my teachers had been born and educated in Europe. It didn't feel right to distrust my teachers. My parents had entrusted me to them. Among the educated European and Indo communities, they were considered to be the best teachers in all of the Netherlands Indies. So I was obliged to trust them.

This science and learning, which I had been taught at school and which I saw manifested in life all around me, meant that I was rather different from the general run of my countrymen. I don't

16

know. And that's how it was that I, a Javanese, liked to make notes—because of my European training. One day the notes would be of use to me, as they are now.

One of the products of science at which I never stopped marveling was printing, especially zincography. Imagine, people can reproduce tens of thousands of copies of any photograph in just one day: pictures of landscapes, important people, new machines, American skyscrapers. Now I could see for myself everything from all over the world upon these printed sheets of paper. How deprived had the generation before me been—a generation that had been satisfied with the accumulation of its own footsteps in the lanes of its villages. I was truly grateful to all those people who had worked so tirelessly to give birth to these new wonders. Five years ago there were no printed pictures, only block and lithographic prints, which gave very poor representations of reality.

Reports from Europe and America brought word of the latest discoveries. Their awesomeness rivaled the magical powers of the gods and knights, my ancestors in the wayang shadow puppet theater. Trains—carriages without horses, without cattle, without buffalo—had been witnessed now for over ten years by my countrymen. And astonishment remains in their hearts even today. The distance from Betawi to Surabaya can be traveled in only three days! And they're predicting it will soon take only a day and a night! A day and a night! A long train of carriages as big as houses, full of goods, and people too, all pulled by water power alone. If I had ever been so lucky to meet Stephenson, I would have made him an offering of a wreath of flowers, all orchids. A network of railway tracks splintered my island, Java. The trains' billowing smoke colored the sky of my homeland with black lines, which faded into nothingness. It was as if the world no longer knew distance—it too had been abolished by the telegraph. Power was no longer the monopoly of the elephant and the rhinoceros. They had been replaced by small manmade things: nuts, screws, and bolts.

And over there in Europe, people had begun making even smaller machines, with even greater power, or at least with the same power as steam engines. Indeed, not with steam—with oil. There were also vague reports saying that a German had made a vehicle that worked by electricity. Oh Allah, and I couldn't really understand what electricity was.

The forces of nature were beginning to be changed by man and put to his service. People were even planning to fly like the shadow puppet character Gatotkaca, like Icarus. One of my teachers had said: Just a little while longer, just a little while, and people will no longer have to force their bones and squeeze out their sweat for so little result. Machines will replace all and every kind of work. People will have nothing to do except enjoy themselves. You are fortunate indeed, my students, he said, to be able to witness the beginning of the modern era here in the Indies.

Modern! How quickly that word had surged forward and multiplied itself like bacteria throughout the world. (At least, that is what people were saying.) So allow me also to use this word, though I still don't fully understand its meaning.

In short, in this modern era tens of thousands of copies of any photo could be reproduced each day. And the important thing was there was one of these that I looked at more often than any other: a photo of a beautiful maiden, rich, powerful, glorious, one who possessed everything, the beloved of the gods.

The rumors, whispered furtively among my school friends, were that even the richest bankers in the world had no chance of courting her. Handsome and manly nobility scrambled head over heels just to be noticed by her. Just to be noticed!

Whenever I had nothing to do, I would gaze at her face while supposing how it would be to court her. How would it be! And how high, too, was her station. And how far away she was, nearly twenty thousand kilometers from where I was: Surabaya. One month's sail by boat across two oceans, five straits, and through one canal. Even then there'd be no certainty of being able to meet her. I didn't dare speak my feelings to a single soul. They would have laughed at me and called me mad.

At the post offices, so rumor was also whispered, letters were occasionally received proposing marriage to this maiden who lived so far away and so high above. None ever reached her. Even if I had been crazy enough to try, it would have been just the same: The post officials would have only kept the letter for themselves.

And that beloved of the gods was the same age as me: eighteen. We were both born in the same year: 1880. Only one figure shaped like a stick, the others roundish, like miscast marbles. The day and the month were also the same: 31 August. If there were any differences, they were only the hour and sex. My parents

18

never noted down the time of my birth. And I didn't know the hour of her birth. As for difference in sex, I was a male, she was a female. And that bewildering difference in time: When my island was blanketed in the darkness of night, her land was lit with sunshine. When her country was embraced by the blackness of night, my island shone brightly under the equatorial sun.

My teacher, Magda Peters, forbade us to believe in astrology. It was nonsense, she said. Thomas Aquinas, she said, once saw two people who were born in the same year, in the same month, on the same day and at the same hour, even in the same place. The joke played by astrology was that one became a great landowner and the other his slave.

Indeed I don't believe in astrology. How could anyone believe in it? It has never lit the way for progress in science and in learning. And it demands of us that we submit to its predictions. There is nothing else we can do except to throw it into the pig's slops bucket. Once I had my fortune told, just for fun. My horoscope was turned over and over. The fortune-teller opened her mouth. She had two gold teeth: If sir is patient, she said, he will surely meet the maiden. So I just prefer to trust my intellect. Even with the patience of all mankind, I would never meet her.

I put my trust in scientific understanding and in reason. With these, at least, there are certainties that can be grasped.

Without knocking on the real door of my rented room, Robert Suurhof—I won't use his real name here—entered. He found me crouched over the picture of that maiden, that beloved of the gods. He burst out laughing; my eyes grew moist, I was so embarrassed. His shout was even more impudent.

"Oho, you philogynist, lady-killer, crocodile! What is the good of wishing for the moon?"

I could have thrown him out. But instead: "Oh . . . you never know!"

Let me tell you about Robert Suurhof: he was then my school friend from H.B.S. (the prestigious Dutch-language senior high school), H.B.S. Street, Surabaya. He was taller than me. In his body ran some Native blood. Who knows how many drops or clots.

"Forget her," he said. His voice had a coaxing, groaning note in it. Then: "There is a goddess here too in Surabaya—beautiful beyond comparison, easily equal to this picture. It's only a picture anyway."

And he mocked me, the one who had defined beauty, by quoting my definition back at me: "Bone structure and body proportion must be in balance. And with fine, soft skin. She must have eyes that shine and lips that are *clever at whispering*."

"You've added '*clever at whispering*,'" I said.

"Yes, then if she curses you, you won't hear."

I offered him silence.

He gave me a look. "If you are a real man, a true philogynist, come with me there. I want to see what you do, whether you're indeed as manly as you say you are."

"I've still got a lot of work to do."

"You're afraid even to descend into the arena," he accused.

That offended me. I knew that the H.B.S. brain inside the head of Robert Suurhof was only clever at insulting, belittling, disparaging, and working evil on people. He thought he knew my weakness: I had no European blood in my body.

"It's on!" I answered. That was several weeks ago, at the beginning of the new school year.

And now all of Java was celebrating, perhaps also the whole of the Netherlands Indies. The tricolor fluttered joyously everywhere: That one-and-only maiden of the photograph, goddess of beauty, beloved of the gods, was now ascending the throne. She now was my queen. I was her subject. Exactly like Miss Magda Peters's story of Thomas Aquinas. She was Her Majesty Wilhelmina. Date, month, and year of birth had given the astrologer the opportunity to raise her to become a queen and to cast me down to become her subject. And my queen would never know that I had walked this earth.

The date was September 7, 1898. Friday. This was in the Indies. Over there in Holland: September 6, 1898. Thursday.

All the school had gone crazy celebrating the coronation: competitions, performances, exhibitions of all those skills and abilities studied by Europeans—soccer, acrobatics, and softball. And none of this interested me. I didn't like sports.

The world around me was bustling. The cannons were booming. There were parades and hymns of praise, but my heart was empty, tormented. So I went, as usual, to my next-door neighbor and business partner, Jean Marais. Jean was a Frenchman and had only one leg. But his story comes later. He greeted me in French, forcing me to use his language.

"*Ça va,* Jean, I have some work for you. One sitting-room suite." I gave him a drawing of what the customer wanted.

"Master Minke!" came a call from next door.

Sticking my head out the window I saw Mrs. Telinga waving to me.

"Jean, I'm going. She may be serving cake."

At home I found no cake. Only Robert Suurhof.

"Ayoh!," he said. "We'll go now."

A new model buggy was waiting for us at the front gate. We climbed aboard; the horses began to move. The coachman was an old Javanese.

"The rent for this must surely be more expensive than for any other," I said in Dutch.

"No fooling, Minke, this is no ordinary buggy, no cheap kretek. It's got springs—perhaps the first in Surabaya. Its springs probably cost more than the rest of the buggy put together."

"I can believe you, Rob. Come on, tell me, where are we going?"

He replied in his insolent, mysterious way:

"A place to which every youth dreams of receiving an invitation, because of the angel that lives there, Minke. Listen, I've had the good luck to be invited by her older brother. Nobody has ever got an invitation, except this one." He pointed to himself with his thumb. "Listen, coincidentally her brother is also called Robert."

"There are a lot of children called Robert now." He took no notice of me and continued.

"We met at a soccer match. And now I am invited to lunch to eat bull calves. That is what interests me most." He glanced slyly at me.

"Bull calves?" I did not understand.

"Veal, to eat veal. That's my problem. Your problem"—he made a noise with his lips, his eyes sharply examining mine—"is that little sister of Robert's. I want to see how far this masculine charm of yours gets you, you philogynist."

The steel frames of the buggy's wheels rattled on as it ground along the stone road of Kranggan Street to Blauran, in the direction of Wonokromo.

"Come on, sing *veni, vidi, vici*—I came, I saw, I conquered." He prompted me to join in between the rattle of the wheels. "Ha-ha, you've gone pale now. He no longer believes in his own virility. Ha!"

"Why don't you take it all for yourself. Veal and this goddess?"

"I? For me—only a goddess with Pure European blood!" So the goddess we were about to visit was an Indo girl, a Mixed-Blood, Indisch. Robert Suurhof—I remind you once again, I'm not using his real name—was also an Indo. When his mother, an Indo, was about to give birth, his father, also an Indo, rushed her to Perak Harbor, boarded the ship *Van Heemskerck,* which was tied up in port, so she had the child there, and he not only became a Dutch subject but a Dutch citizen as well. So he thought anyway. But I found out later that to be born on board a Dutch ship had no legal consequences whatever. Perhaps his behavior was similar to that of the Jews with Roman citizenship. He held himself to be different from his own brothers and sisters. He did not look upon himself as an Indo. If he had been born only one kilometer from that ship, maybe on the docks of Perak, perhaps on a Madurese sampan, and obtained Madurese citizenship, his behavior would have been a bit different. At least I began to understand why he carried on about not being interested in Indo girls. Under the illusion he was actually a Dutch citizen he strove to act as one for the sake of his grandchildren's future. He hoped that, in the future, he'd have a position and salary higher than that of an Indo, let alone a Native.

It was a very beautiful morning. The blue sky was clear, cloudless. Young life breathed nothing but pleasure. I was succeeding in all that I was doing. I was doing well in my studies. And I had an unworried heart and clear emotions. And she who had ascended the throne? That was all over for me. All the decorations on the buildings and gateways were for her. All the official gatherings were also for her. Beloved of the gods! Heavenly goddess! And now Suurhof wanted to make fun of me in front of this other earthly girl whom he also wanted me to conquer.

I did not even notice all the village people walking to town. The yellow stone road went straight to Wonokromo. Houses, dry fields, wet paddy fields, trees enclosed in bamboo lattice along the road, clumps of forest washed with silver rays of sunshine, all of it flew past brightly. And far away in the distance, indistinctly visible, were the mountains, standing silent in their arrogance, like reclining ascetics turned to stone.

"So we're off to a party in clothes like these?"

"No, I just told you. I'm only going to eat, you to conquer."

"Where are we going?"

"Direct to target."

"Rob?" I boxed his shoulder because of my curiosity. "Come on, tell me."

And still he would not say.

"Don't make such a sour face! If you prove your virility,"—he smacked his lips—"I will respect you more than I do my own teacher. If you fail, look out, all your life you will be the butt of my jokes. Remember that well, Minke."

"You're mocking me."

"No. One day, Minke, you'll become a bupati. Perhaps you'll get a regency where the land is arid. I'll pray that you get a fertile one. If this goddess were to be beside you as your raden ayu, all the bupatis of Java would be in a high fever because of their envy."

"Who said I shall become a bupati?"

"Me. And I shall continue my education in Holland. I shall become an engineer. Then we'll meet again. I shall visit you with my wife. Do you know what will be the first question I ask you?"

"You're dreaming. I will never become a bupati."

"Listen, first I will ask: Hey, philogynist, lady-killer, crocodile, where is your harem?"

"It seems you still look upon me as an uncivilized Javanese."

"What Javanese, even a bupati, is not but a crocodile on land?"

"I'm not going to be a bupati."

He laughed at me scornfully. And the buggy still didn't stop, and with time we moved farther and farther away from Surabaya. I had been offended. Actually, I was too easily offended, and my feelings too easily hurt. Rob did not care. Indeed he had once said: The only way a wealthy and powerful Javanese could prove that he did not intend to have a harem was for him to marry a European, Pure or Eurasian. Then there could never be any co-wives or concubines.

The buggy entered Wonokromo district.

"Look to the left," Rob suggested.

I saw a Chinese-style house with a big yard, well kept and with a hedge. The front doors and windows were closed. It was painted red all over. I didn't think it was at all attractive. And we all knew whose house it was and what it was—a pleasure-house, a brothel, owned by Babah Ah Tjong.

But the buggy kept on going.

"Keep looking to the left."

For about one hundred and fifty meters past the pleasure-house the land was empty. Then there stood a two-storied timber house, also with extensive grounds. Standing behind the wooden fence was a big sign with the words Boerderij Buitenzorg— Buitenzorg Agricultural Company.

Everyone who lives in Surabaya and Wonokromo, I thought, knew that was the house of the wealthy Mr. Mellema—Herman Mellema. Everyone thought of that house as Mellema's private palace, even if it was only made of teak. Its grey, wooden-shingle roof was already visible from quite a distance away. Its doors and windows stood wide open—not like Ah Tjong's pleasure-house. There was no veranda. In its place there was a broad, expansive awning overhanging the wooden stairs, which were also wide, wider than the front door.

But that's all that anyone knew, his name: Mr. Mellema. People would see him once or twice only, or once and then never again. But everyone talked about his concubine: Nyai Ontosoroh. People admired her very much. She was handsome, in her thirties, and she managed the whole of this great agricultural firm. People called her Ontosoroh, a Javanese pronunciation of Buitenzorg.

The family and its business were guarded by a Madurese fighter, Darsam, and his men. No one dared to call on that timber palace.

I sat up, startled.

The buggy suddenly turned, passed through the gate, passed the Boerderij Buitenzorg sign, and headed directly to the house's front steps. I shuddered. Darsam, whom I had never seen, appeared in my mind's eye. Just a mustache, nothing but a mustache, a fist, and a giant sickle. I had never heard of anyone receiving an invitation from this eerie and sinister palace.

"Here?"

Robert just spat.

An Indo-Eurasian youth opened the glass door and came

down the steps to greet Suurhof. He appeared to be about my age. He looked European, except he had brown skin. He was tall, well built, sturdy.

"Hi, Rob!"

"Oho, Rob!" greeted Suurhof. "I've brought my friend. It's okay, isn't it? You don't mind, do you?"

He didn't greet me. I was just a Native. He looked at me piercingly. I started to become anxious. I knew we were beginning a new round in a game. If he refused to receive me, Suurhof would laugh and wait for me to crawl back to the main road, driven away by Darsam. He hadn't yet refused, hadn't yet expelled me. With just one movement of his lips, I could be driven out—God! Where must I hide my face? But no, suddenly he smiled and held out his hand.

"Robert Mellema," he introduced himself.

"Minke," I responded.

He still held my hand, waiting for me to give my family name. He raised his eyebrows. I understood: He thought I was an Indo who was not, or not yet, legally acknowledged by my father. Without a family name, an Indo is considered beneath contempt, like a Native. And I am indeed a Native. But no, he didn't demand my family name.

"Pleased to meet you. Come on in."

We went up the steps. His sharp glance did nothing to dispel my suspicions.

But suddenly a new mood replaced suspicion. In front of us stood a girl, white-skinned, refined, European face, hair and eyes of a Native. And those eyes, those shining eyes! ("Like a pair of morning stars," I called them in my notes.) If this was the girl Suurhof meant, he was right: Not only could she rival the queen, she triumphed over her. And she was alive, flesh and blood, not just a picture.

"Annelies Mellema." She held out her hand to me, then to Suurhof.

The voice that came from her lips left an impression that I will remember for the rest of my life.

The four of us sat on a rattan settee. Robert Suurhof and Robert Mellema were soon engrossed in talk about soccer. I felt too awkward to join in. I had never liked soccer. My eyes began to poke around the big drawing room: the furniture; the ceiling;

25

the dangling crystal candle chandelier; the hanging gaslights with their copper piping (I couldn't work out where the main gas tank was); a picture of Queen Emma, who had just abdicated, hanging on the wall in a heavy wooden frame. Being a part-time trader in furniture, just one look at these objects told me that they were nothing but the most expensive, made by master craftsmen. The carpet under the settee was decorated with a motif I'd never come across before. And for the umpteenth time my gaze ended resting on Annelies's face.

"Why are you so quiet?" Annelies asked. She addressed me in familiar Dutch.

Once again I gazed at her face. I hardly dared look into her eyes. Surely she would be repulsed by me. I had no family name and I was a Native too. All I could do was smile—and once again I forced myself to look away toward the furniture.

"Everything is so beautiful here," I said.

"You like it here?"

"Very much," and once again I looked at her.

Even in the middle of all this sumptuousness she appeared grand, a part of it but outshining all these rich and beautiful things.

"Why do you hide your family name?" she asked.

"I haven't hidden it," I answered, and I began to become anxious again. "Do I really need tell?" I glanced over at Robert Suurhof. Before I could look away, he let fly his own glance.

"Of course you do," Annelies said. "Otherwise people will think you're not acknowledged by your father."

"I don't have a family name. Truly I have none," I answered.

"Oh!" she exclaimed slowly. "Forgive me." She was silent for a moment. "That's quite all right," she then said.

"I'm not an Indo," I added in a defensive tone.

"Oh!" she exclaimed once again. "No?"

It felt as if a drum were pounding in my heart. So she knew: I was a Native. I could be thrown out at any moment. I could feel the glances of Robert Suurhof examining those parts of my body that were not covered up. Yes, like a vulture examining a candidate carcass. When I looked up I saw Robert Mellema stabbing at Annelies with his eyes. At that moment he turned to me, his lips becoming a thin, straight line. Oh Lord, what will happen to me? Must I be thrown out like a dog from this beautiful house, accompanied by the cascading laughter of Robert Suurhof? His eyes were knifing at my neck. The Mellema boy hadn't even blinked.

For a moment my vision blurred. All I could see was Annelies's white gown, without a face, without limbs.

And then I began to realize: It had been Suurhof's intention all along to humiliate me here in someone else's house. And now all I could do was to wait for the expulsion.

A moment more and Darsam, the fighter, would be called and ordered to throw me out onto the street.

All of a sudden I heard the shrill laughter of Annelies and this crazed heart of mine felt as if it no longer beat. Slowly I lifted my eyes toward her. Her teeth gleamed, visible, more beautifully white than any I had ever seen. Oh, philogynist! Even in a situation like this you can still admire and praise beauty.

"It's all right to be a Native," she said, still laughing.

Now Robert Mellema's look was directed at his little sister. Annelies, challenging him, looked him straight in the face. Her brother looked away.

What sort of drama was all this? Robert Suurhof did not say anything. Neither did Robert Mellema. Were the two youths in league to force me to apologize? Only because I had no family name and was a Native as well. Why should I? I would not.

"Being Native is good too," Annelies said earnestly. "My mother is a Native. Native Javanese. You are my guest, Minke." Her voice had the tone of an order.

Only then could I breathe freely again.

"Thank you."

"It seems that you don't like soccer. I don't either. Let's sit somewhere else." She stood up and showed me the way, putting out her hand in that sweet, spoiled way of hers. She wanted me to take her by the hand.

I stood up, and excused myself, nodding to her brother and Suurhof. Their eyes followed us. Annelies glanced back with an apologetic smile to the guest she left behind.

As we crossed that broad drawing room, my knees almost gave way. I could feel the glances of the two youths stabbing into my back. We went into the back parlor, which was even more sumptuously furnished.

Here, too, all the walls were made from light brown varnished teak. In the corner there was a dining suite, consisting of one table and six chairs. Close by there were stairs leading up.

Small tables stood on guard, night and day, in each of the other three corners. A vase of European porcelain stood upon each one.

Seeing my eyes fixed upon the display cabinet, she took me over to it. The cabinet stood against the wall opposite the dining table. In it were displayed art objects—I'd never seen such things before.

"I am not carrying the keys with me," said Annelies. "That's the one I like best." She pointed to a small bronze statue. "Mama says it's an Egyptian empress." She thought for a moment. "If I'm not mistaken her name is Nefertiti."

Whatever the name of the figurine, I was amazed that a Native, and a concubine at that, knew the name of an Egyptian empress.

There was also a Balinese carving of the East Javanese king Erlangga, riding on the back of the mythical garuda bird. Unlike the others, it was not made from sawoh wood, but from some other kind that I had never come across before.

On the first shelf there was a row of little ceramic masks, picturing all sorts of animal faces.

"These are the masks from the story of *Sie You Chie*," she explained. "Have you heard the story?"

"Not yet."

"One day I'll tell it to you. Would you like that?"

The question sounded so inviting it drowned out all the sumptuousness, and the differences that existed between us.

"Very much so."

"Then you'd definitely like to come here again."

"An honor."

There were no great clamshells sitting at the feet of the small tables such as I had seen in the bupati buildings. There was a phonograph on a low table with a small wheel on each of its four legs. The lower section of the phonograph table was used as a place to store music. The table itself was ornately carved. It must have been made to order.

"Why are you silent?" she asked again. "You're still at school?"

"A school friend of Robert Suurhof."

"It looks like my brother is really proud to have him as a friend, an H.B.S. student. Now I too have a friend who is an H.B.S. student. You!" Suddenly she turned round toward the back door and called: "Mama! Over here! Mama, we have a guest."

A Native woman entered, wearing a traditional Javanese wrap skirt and a white blouse embellished with expensive lace, perhaps the famous Dutch lace made in Naarden, which we had been told about in E.L.S. She was wearing black velvet slippers embroidered with silver thread. Her neat attire, her clear face, her motherly smile, and her very simple adornments made a deep impression on me. She looked lovely and young; her skin was smooth and light-colored like the langsat fruit. So this was what she looked like, this Nyai Ontosoroh who was talked about by so many people, whose name was on the lips of everyone in Wonokromo and Surabaya, the nyai in control of the Boerderij Buitenzorg.

"Yes, Annelies, who is your guest?" I was even more startled, for she spoke in Dutch.

"An H.B.S. student, Mama."

"Oh yes? Is that so?" Nyai asked me. Her Dutch was good, with correct school pronunciation.

I hesitated. Should I offer my hand as to a European woman, or should I treat her as a Native woman and ignore her? But it was she who first offered her hand. I was dumbfounded and clumsily accepted her grip. This was not Native custom but European. If that's how they do things here, then I, of course, will offer mine first.

"Annelies's guests are my guests too," she said. Her Dutch was so fluent. "How shall I call you? Sir? Sinyo? But you're not Indo."

"Not Indo . . ." What should I call her, Nyai or Madam?

"Are you really an H.B.S. student?" she asked, smiling affably.

"Yes, really."

"People call me Nyai Ontosoroh. They can't pronounce Buitenzorg. Sinyo appears to hesitate to call me that. Everyone does. Don't be reluctant."

I didn't answer. And it seemed she forgave my awkwardness.

"If Sinyo is an H.B.S. student, Sinyo is no doubt the son of a bupati. Bupati of what regency, Nyo?"

"No, Ny, Ny . . ."

"Sinyo is so reluctant to call me by my name. Well, then, call me Mama, like Annelies—that is if you don't feel insulted, Sinyo."

"Yes, Minke," the daughter added. "Mama's right. Call her Mama."

"I'm not the son of any bupati, Mama," and with the use of the new name, my awkwardness, the differences between her and me, even her strangeness, abruptly disappeared.

"Then you must be the son of a patih," Nyai Ontosoroh continued.

"Not the son of a patih either, Mama."

"Very well. But I am so pleased that Annelies has a friend to visit her. Ann, look after him properly, this guest of yours."

"Of course, Mama," she answered cheerfully, now that she had her mother's blessing.

Nyai Ontosoroh left us. I was amazed not only that this Native woman could speak Dutch so well, but also that she was so relaxed with a male guest. Where was she educated? And why was she only a nyai, a concubine? And who educated her to be so free, just like a European woman? What had been a sinister, eerie place was changing into a castle of puzzles.

"I'm glad I have a guest." Annelies became even more cheerful, knowing her mother had no objections. "No one has ever visited me. People are afraid to come here. Even my old school friends."

"Where did you go to school?"

"E.L.S. I didn't finish. I didn't even get to fourth class."

"Why didn't you go on?"

Annelies bit her finger and looked at me.

"There was an accident," she answered, and did not go on. Suddenly she asked: "You're Moslem?"

"Why?"

"So that you don't eat pork."

I nodded.

A maid served chocolate milk and cakes. And the servant didn't come cringing in as she would have before Native masters. She walked in and stood gaping at me in amazement. That would never be allowed by a Native master: A servant must bow down, bow and scrape continuously. How beautiful life is when one doesn't have to cringe before others.

"My guest is Moslem," said Annelies in Javanese to her servant. "Tell them out back not to let pork touch the other food."

30

Then she quickly turned to me and asked, "Why are you so silent?"

"Don't you know?" I asked in return. "Because I never dreamed I'd ever come face to face with such a beautiful goddess as this." She was silent and stared at me with her day-star eyes. I regretted having said it. Hesitantly and slowly she asked, "Who do you mean by this goddess?"

"You," I whispered, and the look on her face changed. She tilted her head. Her eyes opened wide.

"Me? You're saying I'm beautiful?"

I became more daring, insisting, "Without rival."

"Mama!" shouted Annelies and turned around to the back door. Disaster! I exclaimed no less loudly than her—but in my heart, of course. The girl went to the back door. She was going to take the matter to Nyai. Crazy child! Such a contrast with her beauty. And she's going to complain: Minke's being impertinent. Indeed this house is a place of misfortune. No, no, not misfortune. Whatever happens now, it will all be of my own doing.

Nyai appeared at the door and walked toward me.

My heart started to pound again. Perhaps I had done wrong. Punish this impudent one, but don't shame me in front of Robert Suurhof.

"What's the matter now, Ann? Has she started an argument, Nyo?"

"No, we weren't arguing," the girl flashed, then she complained in that sweet, spoiled manner of hers, "Mama"—her hand pointed to me—"imagine, Mama, how could Minke say I was beautiful?"

Nyai stared at me. Her head was tilted a little. She looked at her daughter. In a whisper, with her two hands placed on Annelies's shoulders, she said:

"You know I've often said that you're beautiful? And extraordinarily beautiful? There's no doubt you're beautiful, Ann. Sinyo is not wrong."

"Oh, Mama!" Annelies's face reddened.

Now Nyai sat down on the chair beside me. She said quickly:

"I'm glad you've come, Nyo. She's never mixed properly like other Indo children. She hasn't become an Indo, Nyo."

"I'm not an Indo," the girl contradicted. "I don't want to be an Indo. I only want to be like Mama."

31

I was even more amazed. What was going on in this family?

"Nyo, you heard it for yourself: She'd rather be a Native. Why is Sinyo silent? Perhaps you're offended I'm only calling you Nyo or Sinyo? Without any title?"

"No, Mama, no," I answered hastily.

"You look confused."

Who wouldn't have been confused? Nyai Ontosoroh was behaving as if I were someone who had known her for a long time, but who had forgotten her. It was as if she had given birth to me herself and was closer to me than Mother, even though she looked younger than Mother.

I waited expectantly for Nyai's anger to explode because of my crazy compliments. But she was not angry. Exactly like Mother, who also had never been angry with me. My soul's ear heard a warning too: Beware, don't equate her with Mother. She is just a nyai, living in sin, giving birth to illegitimate children, low in moral character, selling honor to live easily and in luxury. And I couldn't say she was ignorant. Her Dutch was fluent, and polite. Her attitude toward her daughter was refined and wise and open, not like that of Native mothers. She behaved just like an educated European woman.

"The trouble is, Ann," Nyai added, "you don't mix, you only want to be close to Mama." But suddenly, her words were then directed at me. "Nyo, do you usually compliment girls in this way?"

The question flashed at me like lightning. Seeing this as a good omen, I was encouraged to parry like lightning, but carefully.

"If a girl is indeed beautiful, there isn't anything bad in saying so, is there?"

"A European or a Native girl?"

"How is it possible to compliment Native girls? It's impossible even to get close, Mama. European girls, of course."

"Does Sinyo dare do such a thing?"

"We're taught to state our feelings honestly."

"So you are game enough to compliment European girls to their faces?"

"Yes, Mama, my teachers teach European civilization."

"How do they respond to your compliments? Abuse?"

"No, Mama. There is nobody who doesn't like being com-

32

plimented, my teacher says. If someone gets insulted because of a compliment, they say, it's a sign of a dishonest heart."

"So how do these European girls answer?"

"Their answer, Mama, is thank you."

"Like in the books?"

She reads European books, this nyai.

"Ann," she continued, "answer thank you."

Just like a Native girl, Annelies blushed with embarrassment. She didn't say anything.

"And how about Indo girls?" asked Nyai.

"If they've received a good European education, they behave just the same, Mama."

"If they haven't?"

"If not, and especially if they're in a bad mood, they abuse you."

"Sinyo is often abused?"

I knew then that I was blushing. She smiled and turned to her daughter:

"You heard it yourself, Ann. Come on, say thank you. Wait a minute. Nyo, say it again, this compliment of yours, so I can hear it too."

Now I became really embarrassed. What sort of person was I dealing with? She was so clever at capturing and seizing my mind in her hands.

"I'm not allowed to hear?" she asked, looking at my face. "All right."

Again she left us. Annelies and I followed her with our eyes until she disappeared behind the door. And we gazed at each other like two children, equally startled. I burst into uncontrollable laughter. She bit her lip and looked away.

What sort of family was this? Robert Mellema with his frightening, stabbing glances. Annelies Mellema, so childlike. Nyai Ontosoroh, so clever at capturing and seizing control of people's minds that even I lost my judgment and forgot that she was only a concubine. And what about Mr. Mellema, owner of all this abundant wealth?

"Where is your father?" I asked.

Annelies frowned.

"You don't need to know. What for? Even I have no desire to know. Even Mama doesn't want to know."

33

"Why?" I asked.

"Do you like to listen to music?"

"Not now."

And so the conversation dragged on until lunch was served. Robert Mellema, Robert Suurhof, Annelies, and I sat surrounding the table. A young servant, female, stood near the door awaiting orders. Suurhof sat beside his friend and every now and then stole a glance at me and Annelies. Mama sat at the head of the table.

There was much more food than we could have eaten. The main dish was veal, a food I tasted then for the first time in my life.

Annelies sat beside me and served me, as if I were some European master or a very respected Indo.

Nyai ate calmly, like a genuine European woman who had graduated from an English boarding school.

I earnestly examined the position of the spoons and forks, the use of the soup ladle and the knives, carving forks, and also the elaborate dinner service. It was all perfect. The white steel knife seemed to have been sharpened not on stone but on a steel grinding wheel, so there were no scratches. From everything I had read, even the position of the napkins and the finger bowls and the position of the glasses in their silver cases could not be faulted.

Robert Suurhof ate greedily as if he had not seen food for the last three days. I was hesitant even though hungry. Annelies hardly ate anything, only because of the attention she paid to serving me and me alone.

When Nyai stopped eating, naturally I did too, and so did Annelies. Robert Suurhof continued eating and seemed to ignore Nyai completely. And I realized I had not heard the woman speak to her son even once.

"Minke," Nyai said, "is it true people can now make ice? Ice that is really cold, as the books say?"

"It's true, Mama, at least according to the newspapers."

Suurhof swallowed, while glaring at me.

"I want to know if the newspaper reports are true."

"It seems everything will be able to be made by man, madam," I answered, though in my heart I was more amazed that somebody could doubt a newspaper report.

"Everything? Impossible," she replied.

The conversation stopped abruptly. Robert Mellema invited

34

his friend to go outside. They stood and left without taking leave of the Native woman, Nyai Ontosoroh.

"Forgive my friend, Mama."

She smiled, nodded to me, stood up, then left too. The servant cleared the table.

"Mama must continue her work in the office," Annelies explained. "After lunch like this, I have work to do, but out at the back."

"What do you do?"

"Come on, join me."

"What about my friend?"

"No need for you to worry yourself. My brother will invite him to go off hunting. He always goes hunting birds or squirrels with his air gun after lunch."

"Why must it be after lunch?"

"The birds and squirrels are also full and sleepy, so they're not so quick. Come on, come along."

I walked behind her like a child following after his mother. And if she hadn't been beautiful, how would that otherwise have been possible? Oh, philogynist!

Passing through the back door, we entered an area containing steel-hooped wooden barrels. On top of the largest one there was a churning machine. The smell of cow's milk filled the room. People worked without making any sound, as if they were dumb. Now and then they wiped their bodies with a piece of cloth. Each wore a white headband. All wore white shirts with the sleeves rolled up to about ten centimeters above their elbows. Not all of them were men. Some were women; you could tell from the batik kains below their white shirts. Women working in a business. Wearing calico shirts too! Village women wearing coats! And not in their own kitchens! Were they wearing breast-cloths too under their calico shirts?

One by one I looked them over. They only paid attention to me for a moment.

Annelies approached them each in turn, and they greeted her, without speaking, just with a sign. That was the first time I knew this beautiful childlike girl was also a supervisor who must be paid heed to by her workers, male and female.

I was dumbfounded to see women leaving their kitchens in their homes, wearing work clothes, seeking a living in someone

35

else's business, mixing with men! Was this also a sign of the modern era in the Indies?

"You're amazed to see women working?"

I nodded.

"They wear the same uniforms here as workers in Holland, but we can only give them calico."

She pulled me by the hand and took me out into an open compound, the area for drying produce. Several people were working, turning over soya beans, shucked corn, peas, and peanuts. As soon as we arrived, they all stopped work and greeted us by nodding and lifting up their hands. They all wore bamboo farmers' hats.

Annelies clapped her hands and held up two fingers to somebody. A moment later a child worker came up with two bamboo hats. Annelies put one on my head, and wore one herself, and we walked several hundred meters along a path laid with river gravel.

"There are big celebrations on at the moment," I said. "Why aren't they given a holiday?"

"They can holiday if they like. Mama and I never holiday. They're day laborers."

Along the path, up in front of us, quite far away, I could see the two Roberts, each with a rifle slung over his shoulder.

"What work is it you do?" I asked.

"Everything except the office work. Mama does that herself."

So Nyai Ontosoroh does office work. What sort of office work can she do?

"Administration?" I asked, groping around.

"Everything. The books, trading, correspondence, banking . . ."

I stopped in my steps. Annelies also. I stared at her in disbelief. She pulled at my hand and we walked on again toward a row of cattle pens. I already could smell the stench of their dung. It was only because a beautiful girl was taking me that I did not run to avoid it; indeed, I even went into the pens themselves. Only once in all my life. Truly.

The row of pens was very long. In each one people were busy looking after the feed and drink for the dairy cows. The smell of cow dung and of rotten grass made the atmosphere fetid, and I had to resist the urge to vomit.

36

Lifting up the edge of her satin dress, Annelies went up to several cows and patted them on the forehead, talked to them in a whisper, even laughed. I observed her from a distance. She had such easy manners entering the pens and talking with the cows, and in a satin gown like that!

Here too there were women workers. They were sweeping, rinsing down the pen floors, and scrubbing them with very long-handled brushes. They all seemed surprised to see me there.

Annelies walked along the shelves, and I walked along opposite her. She stopped. I saw her talk with a worker, and they both glanced at me, as if sharing a secret.

Another worker, stooped over in deference, walked out in front of me carrying two empty zinc buckets. Her face was pretty. Like the others she wore a breast-cloth and kain. She was bare-footed, wet, dirty, with her toes trumpeting outward. Her breasts were firm and full and by themselves attracted attention. She bowed, glanced up at me from under her forehead, and smiled invitingly.

"Greetings, Sinyo!" She addressed me freely, softly and enticingly.

I'd never met a Native girl so free as that, greeting a man she had never met before. She stopped in front of me and asked in Malay:

"Checking up on things, Nyo?"

"Yes," I said.

Suddenly Annelies was behind me. "How many buckets a day are you getting from your cows, Sis Minem?" Now she spoke Javanese.

"The usual, Non," Minem replied in High Javanese.

Annelies appeared impatient.

"I can still collect more milk than any of them," she said when we went outside. "I don't think you like cows. Let's go to the stables, if you like; or to the fields."

I had never been to a field. There is nothing interesting there. Yet I still followed her.

"Or do you like riding?"

"Ride a horse?" I cried. "You ride horses?"

This childlike girl who had never graduated from primary school suddenly stood revealed as a person of extraordinary character: Not only was she such an efficient manager, but she could

37

ride horses and could get more milk from her cows than any of the other workers.

"Of course. How else could you keep check on fields as large as these?"

We came to a field that had just been harvested. Peanuts. The harvest could be seen heaped on the ground everywhere. There were also piles of peanut stems and plants being carted off for cattle fodder.

"The land here is very good; it can produce three metric tons dry weight of peanuts per hectare. If we hadn't proved it ourselves maybe people would never have believed it," Annelies said. "Good land. First-class quality. Profitable. Even the leaves and stems are good for fertilizer and for cattle fodder."

It seemed that she could read my thoughts: Who cares if it's two or five tons a hectare? I heard her voice: "You're not interested. Let's race the horses. Agree?"

Before I could answer, she pulled my hand. I was dragged along as she ran. I could hear her breathing and then panting heavily. She took me into a big, broad shed that contained coaches, carriages, wagons, and buggies. Saddles, with all sorts of stirrups, hung along the walls. Most of the building was empty.

Seeing my amazement at finding a carriage stable as big as a bupati's office, she laughed, then pointed to a carriage adorned with shining brass and with carbide lights.

"Have you ever seen such a beautiful buggy?"

"Never, never," I answered as I approached the vehicle.

Annelies pulled me along again. We entered into a long, wide stable. There were only three horses inside. Now it was the stench of horses permeating the air and assaulting my sense of smell. She approached a gray-colored horse, embraced the animal's neck, and whispered something in its ear, calling it Bawuk.

Bawuk neighed lightly as if laughing in response. Then Annelies stroked the horse's forehead and it grinned, showing its mighty teeth.

Annelies laughed gaily.

Then, in a serious voice, after whispering while embracing Bawuk's neck, she glanced at me, "We have a guest. That's him. His name is Minke. An alias: It is not a Javanese name, nor Islamic, not even Christian, I imagine. An alias. Do you believe his name is Minke?"

Once again the horse neighed in response.

"Nah!" Annelies said, then to me: "She said that of course your name is an alias."

They were plotting. I was their target. And the other two horses joined in neighing, looking at me accusingly with their big, unblinking eyes.

"Let's go outside," I said, but she went over to the other two horses and stroked each of their backs. Only then did she say to me: "Come on."

"You smell of horses," I said.

She only laughed.

"Apparently it doesn't worry you."

"It's not really important," she answered grumpily. "Bawuk has been treated that way ever since she was small. Mama would be angry if I didn't love her. You must be grateful to everything that gives you life, says Mama, even if it's only a horse."

I didn't annoy her again about the stench.

"Why don't you believe my name is Minke?"

Her eyes shone with disbelief, accusing.

It was, of course, not my idea that my name be, or that people should call me, Minke. I too had been amazed by how it had happened. It is a bit of an involved story. It started when I was still at E.L.S. and did not know a word of Dutch. Mr. Ben Rooseboom, my very first teacher, was always cross with me. I could never answer his questions. I always ended up crying. Yet every day a servant escorted me to that hated school.

I was stuck in first class for two years. Mr. Rooseboom remained cross with me and I remained scared of him. But by the time the new school year arrived, my Dutch was somewhat better. My friends had all gone up to second class. I stayed in first class. I was seated between two Dutch girls, who were always making trouble and annoying me. On one occasion, one of the girls who sat beside me, Vera, pinched my thigh as hard as she could, as a way of getting acquainted. I screamed in pain.

Mr. Rooseboom's eyes popped out frighteningly, and he yelled:

"Quiet, you monk . . . *Minke!*"

From that day, everyone in the class called me Minke, the one and only Native. My teachers followed suit. Then my friends from all the other classes. Also from outside school.

I once asked my elder brother, what did Minke mean? He didn't know. He even ordered me to ask Mr. Rooseboom himself. I didn't dare. My grandfather didn't know Dutch. He couldn't even read or write Latin script. He only knew Javanese, written and spoken. His view was that Minke should be my permanent name: It was a sign of respect from a good and wise teacher. So my real name was almost lost.

I always believed that the name meant something unpleasant. The day my teacher spoke that word *Minke,* his eyes popped out like a cow's eyes. His eyebrows jumped off his broad face. And the ruler in his hand fell to the desk. Goodness and wisdom? Far from it.

I could not find the word in the Dutch dictionary.

Then I entered the H.B.S., Surabaya. My teachers there did not know what it meant either. Unlike us Javanese, they would never make a guess based just on feelings. One even quoted to me from some Englishman: What's in a name? (It was a long time before I could remember the Englishman's name.)

Then we began English lessons. Six months passed and I came across a word similar in pronunciation and spelling to my name. I began to think back over it: eyes popping out and eyebrows ready to disengage from his broad face—for sure he was insulting me. And I remember how Mr. Rooseboom hesitated in saying the name. Fearfully I dared to guess: Perhaps he intended to insult me by calling me monkey.

I've never told anyone what I thought, not even Annelies.

"Minke is a good name," said Annelies. Then: "Let's visit the villages. There are four villages on our land. All the family heads work for us."

All along the road the villagers acknowledged us with respect. They called the girl Non or Noni.

"How many hectares do you have?" I asked.

"One hundred and eighty."

One hundred and eighty! I couldn't even imagine how vast that was. And she continued:

"That's the paddy and fields. It doesn't include the forests."

Forests! She owns forests. Crazy. She owns forests! What for?

"For the firewood," she added.

"Perhaps you own swamps too?"

"Yes. There are two small swamps."

40

"What about mountains?" I asked. "Mountains?"

"You're teasing me." She pinched me.

"Volcanoes, no doubt, so you can catch their flames if they erupt, just like the gods."

"*Iiiih!*" She pinched again.

"What's all that growing over there?" I asked, pointing to a marshy area a few meters away.

"Only reeds. Haven't you ever seen that kind of reed?"

"Let's go over there," I said.

"No," she answered firmly and hunched her shoulders. She shuddered visibly.

"You're scared of that place."

She took my hand and hers felt cold. All of a sudden her eyes became nervous and she tried to tear them away as quickly as possible from the marshes. Her lips were pale. I glanced behind. She pulled my hand and whispered nervously:

"Don't pay any attention. Come on, walk a little faster."

We entered another village, then left it and entered yet another. It was the same everywhere: little, naked children playing everywhere, most with snot hanging from their noses. There were also a few who licked it off. In the shady places, women in late pregnancy sat sewing while carrying their youngest children in kain slings. Two or three women sat in a row looking for head lice.

Several women stopped Annelies and wanted to talk with her, asking for help. And this extraordinary girl, like a mother, affably attended to them all.

She loved her horses because they gave life to her and so it was also with her people. She appeared so grand among the villagers, her people. More grand perhaps than the maiden I had so often dreamed of and who now, in great pomp, had taken her place on the throne to govern the Indies, Surinam, the Antilles, and the Netherlands itself. Even Annelies's skin was finer and more radiant. And Annelies could be approached.

As soon as she finished attending to her people's demands, we continued our walk. Vast nature and that clear, cloudless sky enveloped us. It was scorching hot. It was at that moment I whispered to her these words:

"Have you seen a picture of the queen?"

"Naturally. She's beautiful!"

41

"Yes. You're not wrong."

"Why do you ask?"

"You're more beautiful than she."

She stopped walking just to look at me.

"Thank you, Minke," she answered, embarrassed.

The road became hotter and more still. I jumped over a drain just to see if she would jump or not. She picked up her long dress as high as she could and jumped. I caught her hand, pulled her close, and kissed her upon the cheek. She looked startled, her eyes wide open, examining me.

And I kissed her once again. This time I felt how her skin was smooth as velvet.

"The most beautiful girl I have ever met," I whispered with all my heart's honesty.

She didn't answer, and she didn't say thank you either. She signaled we should go home. She walked along silently all the way. Then a premonition came upon me: You are going to get nothing but trouble from these actions of yours, Minke. If she complains to Darsam, you'll be beaten up before you can even bark.

She walked with her head bowed. I realized only then that her sandals had been left on the other side of the drain. And soon I felt ashamed for pretending not to have noticed.

"Your sandals have been left behind, Ann."

She didn't care. Didn't answer. Didn't look back. She increased her pace.

Quickly I moved up beside her.

"Are you angry, Ann? Angry at me?"

She continued to keep her silence.

The timber palace was visible far away high above the roofs of the other buildings. I could see Nyai watching us from an upstairs window. Annelies, who was walking with her head down, did not know the eyes upstairs followed us until the roofs of the factory buildings blocked Nyai's view.

We entered the house and sat once again on the front-parlor settee. Annelies sat quietly, leaving frozen all my questions. All of a sudden she leaped up and went into another room. As I sat there, I became more and more anxious. She is going to complain to Nyai for sure. I'll get my just desserts now. But no, I will not run.

It wasn't long after that she came out again, carrying a big paper bundle. She placed the object upon the table and said coldly:

"It's late. Rest. That door"—she pointed to the back, at a door—"is your room. In this bundle there are sandals, a towel, and pajamas. You can bathe there. I still have work to do."

Before going, she went to the door she had just pointed out, opened it, and invited me to enter.

Gently she pushed me inside and closed the door from the outside, leaving me alone behind it.

These small and large tensions had made me very tired. My fears about the consequences of my impertinence continued to worry me, though I didn't think I had really done anything wrong. What was it that I had done wrong? Any young man would have behaved the same in the presence of such an extraordinarily beautiful maiden. Didn't my biology teacher say. . . ? Ah, to the devil with biology!

Entering the bathroom was another experience again, another kind of luxury. The walls were lined with mirrors at least three millimeters thick. The floor was made of cream porcelain tiles. I had never seen such a big, clean, and beautiful bathroom. Even a bupati's home would not be equipped with such a bathroom. The bluish water in the porcelain-lined bathtub called out to me to submerge myself in it. And wherever your eyes were directed, it was always yourself that you saw: front, behind, sides, everything.

The bluish clear cool water washed away my anxieties and fears.

And if ever I am rich, I thought, I will build luxury like this. Nothing less than this.

Mama offered me a chair in the parlor. She sat down beside me and tried to start up a discussion about business and trade. It was soon obvious that I knew nothing about these things. She was acquainted with many European terms that I didn't know. Sometimes she would explain them to me just like a teacher. And how clearly this Nyai could explain things!

"Sinyo is interested in business and trade," she said afterwards, as if I had understood everything. "That's very unusual for a Javanese, especially the son of an important official. Or perhaps Sinyo has plans to become a trader or a businessman?"

"I am already trying my hand at business, Mama."

"Sinyo? The son of a bupati? What sort of business?"

"Perhaps also because I'm not the son of a bupati," I replied.

"What business are you in?"

"Top-class furniture, Mama." I began my propaganda. "The latest styles and models from Europe. I go to meet the ships bringing newcomers from Europe. I also visit the houses of the parents of my school friends."

"And Sinyo's progress at school? You're not left behind?"

"Never, Mama."

"Interesting. For me, those who really endeavor are always interesting. Does Sinyo own his own furniture workshop? How many tradesmen?"

"No, I only sell the furniture. I carry pictures with me."

"So you came here to sell furniture? Let's see your pictures."

"No. I came here without bringing anything. But if Mama feels it necessary, I will bring them another time: wardrobes, for example, as in the palaces of Austria or France or England— Renaissance, baroque, rococo, Victorian. . . ."

She listened to me carefully. Twice I heard her smack her lips, I don't know if in praise or as an insult. Then she said slowly:

"Happy are they who eat from the products of their own sweat, obtain pleasure from their own endeavors, and advance because of their own experiences."

The tones sounded as if they had come out of the chest of a priest in a wayang performance. Then she called out:

"Fantastic!" She was looking up at the head of the stairs. "Ah!"

Down those stairs descended the angel Annelies, in a batik kain and a traditional laced kabaya blouse. Her sanggul bun hairstyle was a bit too high, revealing her long white neck. Her neck, arms, ears, and bosom were decorated with a pattern of green-white emeralds, pearls, and diamonds. (Really, I didn't know which were diamonds and which were the others, what was real and what was fake.)

I was entranced. She must have been more beautiful and arresting than Jaka Tarub's angel in the legends of *Babad Tanah Jawi*. She was smiling nervously as if embarrassed. The adornments she was wearing were somewhat, indeed definitely overdone, too extravagant. And I knew she had dressed up for me and me alone.

And for a countenance and presence as beautiful as that, there was no need for any adornment. Naked too, she would remain beautiful. How foolish of us to think that the beauty bestowed by the gods does not always triumph over the inventions of humans. With all those adornments from the sea and the land she looked alien, while the Javanese clothes, which she was not used to wearing, made her movements like those of a wooden doll. Everything about her seemed somehow pretentious. But it didn't matter—what is beautiful stays beautiful. It was up to me to cleverly ignore her extravagances.

"She has dressed up for you, Nyo," whispered Nyai.

Annelies walked up to us while still smiling and perhaps with a thank you readied in her heart. But before I could get in my compliment, Nyai got in first:

"From whom did you learn to dress up and adorn yourself like that?"

"Ah, Mama!" she exclaimed, prodding her mother's shoulder and glancing at me with her big eyes. She was blushing.

I was embarrassed to be listening to such a conversation between mother and daughter: too intimate to be heard by a stranger. Yet near Mama I felt I ought to be resolute. I had to leave behind an impression of being a man who was resolute, interesting, dashing, an unappeased conqueror of the goddess of beauty. In front of the queen I think I would also have had to exhibit the same attitude. That is the cock's plumage, the deer's antlers, the symbol of virility.

I knew what was proper and I did not involve myself in the affairs of mother and daughter.

"See, Ann, Sinyo was ready to go home. It's fortunate we stopped him. Otherwise, he would have really missed out on something!"

"Ah, Mama!" Annelies said again, in her sweet, spoiled manner, and prodded her mother. Her eyes glanced at me.

"Well, what about it, Nyo? Why are you silent? Have you forgotten your own custom?"

"Too beautiful, Mama. What words are appropriate for beauty's beauty?"

"Yes," added Nyai, "fit to become queen of the Indies, isn't she, Nyo?" and she turned to me.

The relationship between mother and daughter seemed strange to me. Maybe it was the result of the illegitimate marriage

45

and birth. Perhaps this is the atmosphere in the homes of all nyais. Perhaps even among modern families in Europe today and among Indies Natives far in the future. Or perhaps it wasn't right, but abnormal. Yet I liked it. And luckily the mutual praising finally ended without having led anywhere.

The light began to fade. Mama talked on. Annelies and I just listened. There were too many new things, which my teachers had never mentioned, that proceeded from her lips. Remarkable. And I was still not allowed to go home, although:

"Dokar?" she said, "Out at the back, there are many such carts. If you like you can even go home in a carriage."

A young boy began to light the gas lamps. I still did not know where the mains were located.

The servants began to prepare the dining table.

The two Roberts were summoned into the back parlor. Dinner began in silence.

Another servant entered the front room, closing the door. The back-parlor light, covered by a milk-white glass shade, shone dimly. No one said a word. Eyes just moved about from plate to bowl, from bowl to dish. Spoons, forks, and knives clinked as they touched the plates.

Nyai lifted up her head. The front door could be heard opening, without any knock, without announcement. I looked up at Nyai. She looked vigilantly toward the front room.

Robert Mellema glanced in the same direction. His eyes shone with pleasure and his lips had a satisfied smile. I also wanted to glance behind, to where their looks were directed. I held back my desire; it wasn't polite, not gentlemanly. So I glanced at Annelies. Her head bowed down, her eyes raised high, clearly she was straining her ears.

Deliberately, I stopped my spoon in midair and focused my hearing on the area behind me. Shoes walking, scraping along the floor. As time passed they became clearer, closer. Nyai stopped eating. Robert Suurhof did not put the food in his mouth; he put the spoon and fork down on his plate. I heard the steps coming closer, drowning out the tick-tock of the pendulum clock.

Robert Mellema continued eating as if nothing was happening.

Finally Annelies, who was sitting beside me, also glanced behind. She blinked open her eyes, startled. Her spoon dropped with a clang to the floor. I tried to pick it up. A servant came running and took it. Then the servant quickly got out of the way. Annelies stood up as if she wanted to confront this new arrival, who was getting closer.

I placed my spoon and fork on the plate and, following Annelies's example, stood up and turned around.

Nyai also stood in readiness.

A shadow, splayed out by the front-room lamps, became longer and longer. The dragging steps became clearer and clearer. Then a European man emerged—tall, big, fat, too fat. His clothes were rumpled and his hair in a mess, who knows if really white or gray.

He looked in our direction. Stopped a moment.

"Your father?" I whispered to Annelies.

"Yes." Almost inaudible.

Looking straight at me Mr. Mellema, dragging his feet, walked towards me. Towards me. He stopped in front of me. His eyebrows were bushy, almost white, and his face was frozen like chalk. For a moment my eyes fell to his shoes, which were dusty, unlaced. Then I remembered what my teachers had taught me: Look those who want to talk to you in the eyes. Quickly, I lifted my eyes and offered my greetings:

"Good evening, Mr. Mellema," in Dutch and in a quite polite tone.

He growled like a cat. His rumpled clothes were loose on his body. His hair, uncombed and thin, covered his forehead and ears.

"Who gave you permission to come here, monkey!" He hissed his sentence in bazaar Malay, awkwardly and, in accord with its contents, crudely.

Behind me Robert Mellema coughed. Then I heard Annelies holding back a sob. Robert Suurhof put his shoes into action and stood up also to extend his greetings. But the ogre in front of me paid him no heed.

I admit it: My body shook, although only a little. In such a situation I could only await words from Nyai. I could expect nothing from anyone else. It was going to be a disaster for me if she stayed silent. And indeed she was silent.

"You think, boy, because you wear European clothes, mix with Europeans, and can speak a little Dutch you then become a European? You're still a monkey."

"Close your mouth!" shouted Nyai loudly in Dutch. "He is my guest."

Mr. Mellema's eyes shifted dully to his concubine. And must something happen because of this uninvited Native?

"Nyai," said Mr. Mellema.

"A mad European is the same as a mad Native!" Her eyes burned with hatred and disgust. "You have no rights in this house. You know where your room is." The nyai pointed to a door. And her pointed finger was clawed.

Mr. Mellema still stood in front of me, hesitant.

"Do I need to call Darsam?" she threatened.

The tall-big-fat man was confused; he growled in answer. He turned and walked, dragging his feet, to a door next to the room I had just occupied, and disappeared behind it.

"Rob," Robert Mellema said to his guest, "let's go outside. It's too hot in here."

They went out together, without excusing themselves to Nyai.

"Trash!" Nyai cursed.

Annelies was sobbing.

"Be quiet, Ann. Forgive us, Minke, Nyo. Sit down again. Don't make a racket, Ann. Sit down in your chair."

We both sat down again. Annelies covered her face with a silk handkerchief. And Nyai still kept an eye on the just-closed door.

"No need to be ashamed in front of Sinyo," Nyai said without looking at us, still in a rage. "And you, Nyo, you may never forget this. I'm not ashamed. Sinyo shouldn't be shocked or feel ashamed either. Don't be angry. I've done exactly what I had to. Just pretend that he doesn't exist, Nyo. Once I was indeed his faithful nyai, his loyal companion. Now he is only worthless garbage. All he is good for now is shaming his own descendants. That is your father, Ann."

Satisfied after her outburst, she sat down again. She didn't resume her dinner. The look on her face was hard and sharp. Calmly, I looked at her. What sort of woman was this?

"If I wasn't hard like that, Nyo—forgive me that I must offer a defense for myself in my humiliation—what would become of all this? His children . . . his business . . . we would be reduced to destitution. So I do not regret acting this way in front of you, Nyo." She lowered her voice as if pleading with me. "Don't think

48

me insolent and rude, Nyo," she said, continuing in her beautiful Dutch. "It is all for his own good. I treat him the way he wants. This is what he wants. It is the Europeans themselves who have taught me to act this way, Minke, the Europeans themselves." Her voice pleaded with me to believe. "Not at school, but in life."

I was silent. I nailed every one of her words into my memory: not in school, in life! Don't think me insolent and rude! Europeans themselves have taught me this. . . .

Nyai stood up, walked slowly towards the window. And behind the door she pulled a cord that ended in a bunch of tassels. In the distance a bell could be heard ringing indistinctly. The servant girl who had just vanished reappeared. Nyai ordered her to take away the food. I still didn't know what I was supposed to do.

"Go home now, Nyo," she said.

"Yes, Mama, it's better I go home."

She walked up to me. Her eyes had their original motherly gentleness.

"Ann," she said still more softly, "let your guest go home now. Wipe away those tears."

"Forgive us, Minke," Annelies whispered, holding her sobs back.

"It's nothing, Ann."

"When holiday time arrives later, come and spend the vacation here, Nyo. Don't hesitate. Nothing will happen. What do you think? Agree? Now Sinyo must go home. Darsam will escort you in the cart."

She walked again to the door and pulled that cord. Then she sat back down in her seat. She was amazing, this nyai: The people and everything around her were indeed in her grip, and I, myself, too. From what school had she graduated that she appeared so educated, intelligent? And she was able to look to the needs of several people at once, with a different manner for each. And if she did graduate from a school, how was she able to accept her situation as a nyai? I couldn't understand any of this.

A Madurese man arrived. He was approaching forty, shirt and pants all black, and an East Javanese destar headband on his head. A short machete was fastened at his waist. His mustache was twirled up high, pitch black and thick.

Nyai gave him an order in Madurese. I didn't catch all of

what it meant. She was probably ordering that I be escorted safely home in a dokar.

Darsam stood straight. He didn't speak. He looked at me with searching eyes—as if he wanted to memorize my face—without blinking.

"The young master is my guest, is Miss Annelies's guest," said Nyai in Javanese. "Take him home. Don't let anything happen on the way. Be careful." Apparently this was only a translation of the earlier Madurese.

Darsam raised his hand without speaking and left.

"Sinyo, Minke," Nyai confided, "Annelies has no friends. She is happy that Sinyo came here. You, of course, don't have a lot of time. I know that. Even so, try to come here often. You don't need worry about Mr. Mellema. I will look after him. If Sinyo would like, we would be very happy for you to live here. You could be taken to school each day by buggy. That's if Sinyo would like."

Such a strange and frightening house and family! It's no wonder they have such a sinister reputation. And I answered:

"Let me think about it first, Mama. Thank you for such a generous invitation."

"Don't refuse us," Annelies said. There was rebuke in her voice.

"Yes, Nyo, think about it. If you have no objections, Annelies will look after it all. Isn't that so, Ann?"

Annelies nodded in agreement.

The carriage could be heard coming along beside the house. We walked to the front of the house and found Robert Suurhof and Robert Mellema sitting silently, looking out at the darkness. The carriage stopped in front of the steps. Suurhof and I went down the steps and boarded the carriage.

"Good night everybody, and thank you very much, Mama, Ann, Rob!" I said.

And the carriage began to move.

"Stop!" ordered Mama. The carriage stopped. "Sinyo Minke! Come down here first."

Like a slave I was caught in her grip. Without stopping to think for a moment, I climbed out and approached the steps. Nyai descended one step and so did Annelies, and Nyai said slowly into my ear:

"Annelies has told me, Nyo—don't be afraid—is it true, you kissed her?"

Even a flash of lightning would not have startled me so greatly. Anxiety crawled through my body, down to my feet, and my feet tripped.

"It is true?" she insisted. Seeing I couldn't answer, she pulled Annelies and drew her to me. Then, "So it's true. Now Minke, kiss Annelies in front of me. So that I may know that my daughter does not lie."

I trembled. Yet I could not resist her command. And I kissed Annelies on the cheek.

"I'm proud, Nyo, that it's you who kissed her. Go home now."

I was unable to say a word all the way home. I felt as if Nyai had cast a spell over my mind. Annelies was indeed gloriously beautiful. Yet her clever mother subdued people so they would bow down to her will.

Robert Suurhof didn't speak either.

And the carriage rattled as it went on grinding the street pebbles. The carriage's carbide light split open the darkness relentlessly. Our carriage was the only one on the road that night. It appeared that everyone had streamed into Surabaya to celebrate the coronation of the maiden Wilhelmina.

Darsam escorted me to my boarding house in Kranggan. He stayed until he saw me enter the house before he left to escort Suurhof home.

"Ai-ai, Master Minke!" Mrs. Telinga, my talkative old landlady, called out. 'So young master doesn't eat at home anymore? I've just put a letter in your room. I see that you still haven't read the earlier letters either. The envelopes haven't even been opened. Remember Young Master, those letters were written, were given stamps, and were sent to be read. Who knows if there may be something important in them? They all seem to come from the town of B——. So, Young Master, what about it? Tomorrow there'll be no shopping money left, eh."

I gave a few coins to the garrulous, good-hearted woman. She said thank you over and over again, as usual, without it needing to come from her heart.

There was hot chocolate ready for me in my room. I drank it

down quickly. I took off my shoes and shirt, jumped onto the bed, and started to reflect upon all that had happened. But my eyes fell upon the portrait of the goddess near the oil lamp on the wall. I got out of bed, studied it well, then turned it over. And I climbed back into bed.

I pushed aside the Surabaya and Batavia papers, which were, as usual, placed on my pillow. It had become my custom to read the papers before sleeping. I don't know why but I liked to seek out reports about Japan. It pleased me to find out that their youth were being sent to England and America to study. You could say I was a Japan-watcher. But now there was something more interesting—that strange and wealthy family: Nyai, with her power to grip people's hearts as if she were a sorceress; Annelies Mellema, who was beautiful, childlike, yet experienced and able in managing workers; Robert Mellema with his sharp glances, who cared about nothing except soccer, not even his own mother; Mr. Mellema, as big as an elephant, sullen, but powerless over his own concubine. Each like a character in a play. What sort of family was this? And myself? I too was powerless before Nyai. Even as I turned over on the bed her voice still called: Annelies has no friends! She is happy Sinyo has come here. You, of course, don't have much time. Even so, try to come here often . . . we would be very happy if you were to stay here. . . .

It felt like I had only been asleep a little while when there was a commotion outside the house. I lit the oil lamp in my room. Five o'clock in the morning.

"There is a package. For Young Master Minke"—I heard a man's voice—"milk, cheese, and butter. There is also a letter from Nyai Ontosoroh herself."

3

Life went on as usual. It was, perhaps, only I who changed. Boerderij Buitenzorg in Wonokromo continued calling, summoning me, every day, every hour. Was I the victim of black magic? I knew many Pure and Indo-European girls. Why was it only Annelies I saw before me? And why did the voice of Nyai not want to go from my soul's ear? Minke, Sinyo Minke, when are you coming?

I was confused.

Every day I left for school with little May Marais. I would walk hand in hand with her as far as her school at E.L.S., Simpang. Then I walked on by myself to my school on H.B.S. Street. I closely observed every carriage driver that passed by me, just in case it was Darsam. And whenever a carriage wanted to pass me from behind, I had to look around. It was as if I had some business with every carriage that passed.

At school, Annelies also hovered before me continuously. And over and over again came Nyai's voice. When are you coming? She has got all dressed up for you. When are you coming?

Robert Suurhof never bothered me with anything about

Wonokromo. He avoided me. He refused to honor his promise to respect my success with Annelies. And I somehow felt as if I were separated from reality by a gray veil. Everything was unclear; all feelings uncertain. All my school friends, Pure European or Indo, male and female, it was as if they had all changed. And they too saw changes in me. I was no longer the same easy-to-get-on-with and affable Minke.

One day, on my way home from school, I went straight to Jean Marais's workshop. He was, as usual, absorbed in his drawings, sketches, or in some design he was preparing. I hadn't wanted to go straight back to my lodgings. I didn't feel like going down to the harbor either. I didn't want to go to the auction-paper office to write up advertising texts. But I had no inclination to write any serious journalism either. I certainly had no desire to go visiting my friends' homes to try to sell furniture or seek orders for portraits.

I didn't feel like doing anything. All my body wanted to do was lie in bed turning over and over while I remembered Annelies. Only Annelies, that childlike maiden.

Mrs. Telinga never tired of asking to hear the story of my visit to Boerderij Buitenzorg, only afterwards to have me listen to her coarse, repetitive insults: "Young Master, Young Master, of course Young Master likes the daughter; but it's her mother who has the great lust. Everybody, of course, says her daughter is beautiful. No one dares go there. Young Master is very lucky. But remember this about Nyai, lest Young Master be gobbled up by her!"

Not only Mrs. Telinga and I knew, but it felt as if the whole world knew, that such indeed was the moral level of the families of nyais: low, dirty, without culture, moved only by lust. They were the families of prostitutes; they were people without character, destined to sink into nothingness, leaving no trace. But did this popular judgment apply to Nyai Ontosoroh? This was what was confusing me. No, she wasn't like that. Or was I just a careless observer? Maybe I just didn't want to know. All social classes had passed judgment on the Nyai, as well as all races: Native, European, Chinese, Arab. How could I, just one person, say no. Her order that I kiss Annelies, wasn't that a sign of her low morals? Perhaps. Yet Mrs. Telinga's insults offended something inside me. Perhaps it was because I was fantasizing. During the last few days I had been trying to convince myself that what had taken place between Annelies and myself was just a normal event in the life of a young man and a young woman. It happens to people from all walks of life: kings, traders, religious leaders, farmers, workers, even the gods in heaven. True. But an invisible finger pointed accusingly at me and said: The trouble is you're trying to justify your own fantasies.

And so that afternoon I found myself compelled to go and ask Jean Marais. I could not yet hope for a really serious conversation with him, although his Malay was getting better every day. He didn't know Dutch. That was the difficulty. His Malay was limited. My French was pretty hopeless. He resisted learning Dutch with all his might, even though he had fought in Aceh with the Dutch Colonial Army for more than four years. His Dutch was confined to military terms.

But he was my oldest friend, my companion in business. It was only proper that I ask him.

The workmen were finishing off furniture ordered by someone called Ah Tjong. I suspected that it was the one who owned the brothel next door to Nyai Ontosoroh's. Because the order was for European-style furniture, the Chinese hadn't gone to a Chinese carpenter. I had received the order through someone else.

Jean was playing with his pencil, making a sketch for his next picture. "I need to disturb you, Jean," I said and sat on the chair at the drawing table. He lifted up his face and looked at me. "Do you know the meaning of *sihir*?"

He shook his head.

"Guna-guna?" I asked.

"Yes—black magic—so I have heard anyway. The Africans practice it, people say. That's if I have heard properly."

I began to tell him of my situation, about being under a spell, and about the popular view of the nyais in general, and of Nyai Ontosoroh in particular.

He put his pencil down on top of the drawing paper, stared at me, and tried to capture and understand each of my words. Then, calmly and in a mixture of several languages:

"You're in trouble, Minke. You've fallen in love."

"No, Jean. I am not in love. She is certainly an attractive, enchanting girl, but in love, no."

"I understand. You're really in trouble; it's serious when you can't tell somebody they've fallen in love. Listen, Minke, your young blood wants to have her for yourself, and you're afraid of what people will say." He laughed slowly. "You must pay heed to and respect what people think if they are correct. If they're wrong, why pay them any heed? You're educated, Minke. An educated person must learn to act justly, beginning, first of all, with his thoughts, then later in his deeds. That is what it means to be educated. Go and visit this family two, or maybe three more times. Then you might be able to judge for yourself if Nyai and her family deserve their bad reputation."

"So you think I should go back?"

"I think you should find out for yourself if what people say is fair or not. To go along with unfair gossip is wrong. You might find you're judging a family that is perhaps better than the judge himself."

"Jean, I've been asked to go and stay there."

"Take up the offer. Only don't forget your studies. But you don't really need to chase after new orders any more. Look, there are still five portraits that have to be finished. And this"—he patted his sketch paper—"I'm going to paint something I've long dreamed of doing."

I looked at the sketching paper in front of him. The picture immediately made me forget my own problems. A Netherlands Indies soldier—it was obvious from his bamboo hat and his sword—was thrusting his foot down onto the stomach of an Acehnese fighter. The soldier was pushing his bayonet down towards the bosom of his victim. The bayonet pressed onto the black shirt, and from under the shirt emerged the breast of a young woman. The eyes of the woman were wide open. Her hair fell in bunches over fallen bamboo leaves. Her left hand was resisting as she tried to rise. Her right hand powerlessly held a dagger. Above them both, like an umbrella, was a cluster of bamboo bent down by the attack of a strong wind. It was as if only those two lived: one who was to kill and one who was to be killed.

"But this is so vicious, Jean. You like to talk about beauty. Where is the beauty in viciousness, Jean?" He sucked on his cigarette. Then:

"It's not easy to explain, Minke. This picture is very personal, not intended for public display. Its beauty is in the memories it records." He was silent, and I realized:

"So you are the soldier, Jean? You've carried out such barbarity as this?" He shook his head. "You killed this young woman?" He shook his head again. "So you freed her?" He nodded. "She must have been grateful to you."

"She was not grateful, Minke. She asked me to kill her. She was ashamed that she had been sullied by the touch of an infidel."

He answered dispiritedly, and as if his words were not directed at me, but at his own past, which was now far beyond his reach.

"And now she is dead. Her younger brother sneaked into the camp and stabbed her in her side with a poison-tipped dagger. She died immediately. Her killer then himself died, hearing his own cries: 'To hell with you infidel, follower of the infidels!' "

"Why did her brother stab her?" Already I'd forgotten all about my own troubles.

"Her brother had continued fighting for his country, for his beliefs. His sister had surrendered. There was nobody there when she died, Minke. Her child was being taken for a walk at the time. Her husband was away on duty."

"So this woman lived amongst the soldiers? As a prisoner? As a prisoner until she had a child?"

"At first she was a prisoner. Then later, no longer," he answered quickly.

"So she married?"

"No. She didn't marry."

"And the child who was out walking, where did it come from?"

"That child was the baby she gave me, my own child, Minke."

"Jean!"

"Minke, don't tell this story to May."

All of a sudden I was overcome by emotion. I ran to find May, who was safely asleep on a wooden divan, without a sheet. I picked her up and I kissed her. She was startled and looked at me with wide-open eyes. She didn't say a word.

"May! May!" I cried to her, to myself, and I carried her out and found Jean Marais again. "Jean, this is your child. This is that baby, Jean. You're not lying to me, Jean? You're lying to me, aren't you?"

The Frenchman, his chin now resting on his hand, was gazing

out of the house into the distance. He didn't want to repeat his story. He didn't want to answer. And I remembered how often he'd spoken of his love for his wife.

"That's why you advise me to go to Wonokromo?" I asked.

"Love is beautiful, Minke, very, very beautiful, but perhaps disaster follows. You must dare to face its consequences."

The child prattled. "Will we go for a walk, Papa?" and when he said they would go after she had bathed, she ran merrily off. I said:

"Jean, I'm not sure I love this Wonokromo girl."

"Perhaps you don't or at least don't yet love this girl. It's not up to me to determine that. And also there is no love that appears suddenly out of the blue, because love is a child of culture, not a stone dropped from heaven. You must test yourself, your own heart. This girl may like you. It's clear from what you've told me that her mother is very fond of you. Fond of you after the first meeting. I don't believe in black magic. Perhaps it exists, but I don't need to believe in it, because it can only be common where life is still at a very simple level of civilization. Moreover you've said Nyai does all types of office work. Someone like that is not going to believe in black magic. She will believe more in strength of character. Only people without any character practice black magic. Nyai knows what she needs. Perhaps she realizes how lonely her daughter is.

"Do you want to take May for a walk this afternoon?"

"You never take her!" I replied. "She wants to go walking with you."

"Not yet, Minke. Take pity on her. People will stare at us both. One day she will hear someone say: Look at the lame, stumped foreigner and his child! No, Minke. She mustn't have her young soul scarred by some unnecessary hurt, especially not one caused by her own father's deformity. She should love me and look upon me as the father who loves her, without regard to the voices and views of others."

I had never heard him talk so much. I had never seen him so depressed. What was happening inside him? Perhaps he was longing for his past, lost and no longer within his reach? Or for the country where he was born, brought up, and saw the sun for the first time? A country that he doesn't dare return to because he's without one leg and has a daughter born in a foreign country? Or is he longing to create a painting that will force his country to acknowledge his greatness as an artist?

"You have never had any time for pity, Jean," I rebuked him.

"You're right, Minke. As I've told you before, pity is the feeling of well-intentioned people who are unable to act. Pity is only a luxury, or a weakness. It is those who are able to carry out their good intentions who deserve praise. I can't, Minke. The more I reflect on it the more beautiful that word *pity* sounds here in these Indies, though not in Europe."

He sounded even more gloomy.

"This is not the Jean I know," I chided him. "I'm worried about you. You don't seem to be yourself, Jean."

"Thanks for your concern, Minke. You're becoming sharper every day."

May came in. As soon as she found out her father wouldn't be taking her, the look on her face changed.

"Go with Uncle Minke, May. It's a pity but I still have work that I must finish. Don't frown like that, darling."

I took the little French-Acehnese Mixed-Blood girl by the hand and left the house.

"Papa never wants to walk with me," she complained in Dutch. "He doesn't believe I'm strong enough to lead him, Uncle. I'm able to make sure he doesn't fall down."

"Of course you're strong enough, May. He'll take you another time."

I took her to Koblen field, and she began to forget her disappointment. We sat on the grass and watched the kites battling each other. She began to prattle in Javanese mixed with Dutch and some French. I didn't pay any attention. I just said yes to everything. My own thoughts were still confused, under attack from several directions: the Mellema family, the Marais family, the changing attitude of my school friends towards me, and my own changing feelings. Some kites broke loose and whirled around the sky aimlessly.

May pulled at my hand, and pointed at a patch of clouds on the horizon.

"You love your father, May?"

She looked at me with amazed eyes. I saw Jean Marais in her face. I could not find the smallest trace of the face of the young woman sprawled out under the bamboo, threatened by the bayonet. This was perhaps what Jean had looked like as a child. And this little Marais does not know at all who her father really is.

Jean, so he said himself, had once studied at the Sorbonne. He never told me in what department or to what level. Commanded by the voice of his own heart, he abandoned lectures and poured all his strength into painting. He lived in the Latin Quarter in Paris and hawked his paintings on the street. His works always sold well, but never caught society's attention or that of the Parisian critics. While hawking his paintings, he also did carvings on the street. Five years passed. He still didn't get anywhere. He was bored with his surroundings—with the mobs of sightseers who watched him make African statues or other carvings; with Paris; with his own society; with Europe. He longed for something new that might fill life's barrenness. He left Europe, went to Morocco, Libya, Algeria, and Egypt. He still didn't find that something he sought but could not identify: He never felt satisfied; he was always anxious and agitated. He still couldn't create the paintings of which he dreamed. He left Africa. By the time he arrived in the Indies, his money had run out. The only road open to him was to join the Dutch Indies Army. He joined, underwent training for several months, and departed for the front in Aceh. In his army unit too he lived within himself, having almost no contact with anyone except through the orders he received in Dutch. And he was unwilling to learn that language.

May Marais did not know, or did not yet know, any of this.

I can paint and be a soldier too, Jean Marais resolved. The Indies Natives are very simple. They will never win any war. How can daggers and spears defeat rifles and cannons? he thought. He was sent to Aceh as a private first class. The commander of his platoon, Corporal Bastian Telinga, was an Indo-European. If Jean hadn't been a Pure, he would certainly never have got higher than private second-class. Jean lived amongst the other Pure Europeans who spoke no Dutch: Swiss, Germans, Swedes, Belgians, Russians, Hungarians, Romanians, Portuguese, Spaniards, Italians— from almost all the nations of Europe—the rubbish discarded from their own countries. They were people who had given up hope, or bandits on the run, or people running from debts, or bankrupt because of gambling and speculation—adventurers all. And none of them was less than a private first-class. Second-class was reserved for Indos and Natives—generally Javanese from Purworejo.

Why mainly Natives from Purworejo? I once asked. They, Jean said, are a very calm people, with very strong nerves. That's why the army chose them to fight the Acehnese, who are as tough and hard as steel, men of action, and able to instill terror in most people. Perseverance and stamina were absolutely essential to survive in Aceh.

He soon had to admit that his initial views about the Natives' ability to wage war turned out to be wrong. The Acehnese had great ability; it was only their weapons that were inadequate. They were also very well organized, although he acknowledged the Dutch army's outstanding ability in selecting personnel.

Jean once admitted to me he had been wrong to say that dagger, spear, and Acehnese bamboo trap would not be able to face up to rifle and cannon. The Acehnese waged war in their own special way. Using the local environment, all their abilities, and their beliefs, they were able to destroy much of the strength of the colonial army. He was astounded by this. They fought to defend what they regarded as their rights, mindless of death. Everybody, even the children! Even when defeated they fought back. They fought back with all their abilities and all their inabilities.

He once told me another story in that mixture of languages of his. The Acehnese defenses had been pushed far into the interior and to the south, to the region of Takinguen. An Acehnese commander, Tjoet Ali, had lost a great many of his men and much of his territory, yet morale was still high, a secret Jean could not fathom. They kept fighting, not only resisting the army, but resisting their own decline as well. The army's communications links were their easy targets: bridges, roads, telegraph lines, trains, and railways. Then came the drinking water, which they poisoned, surprise attacks, bamboo traps, ambushes, stabbings, running amok in the barracks.

The Dutch generals almost gave up. The Dutch were only ever able to destroy the children, the grandmothers and grandfathers, the ill, the pregnant women. And these helpless people, Jean said, felt fortunate if they died at the hands of the army. Jean's superiors spread rumors that the casualties amongst the European soldiers never reached the three thousand that they did in the Java War; but everyone was still afraid with every square of land that they stepped upon.

And Jean Marais began to admire and love these noble, heroic

61

Natives, with their strong character and personality. For twenty-seven years they had waged war, confronting the most powerful weaponry of the age, the product of the science and the experience of the whole of European civilization.

Love is beautiful, Minke, very, very beautiful, he had said. But he never told me how he came to love the enemy he captured or how she may have come to love him, and to give him a beloved child, May, now sitting in my lap prattling.

I stroked her hair. For how many months was your mother able to feed you from her breasts, sweet one? You have lost something that nothing and no one can ever replace.

"Look over there, Uncle," she called out in Dutch, "above those clouds. Kites shouldn't be like crabs!"

"Yes, crabs shouldn't be flying in the sky. The clouds are getting darker, May, let's go home."

Jean Marais was still crouched over his drawing table. He looked up when we entered. May quickly went to him and told him about the crab kites over the clouds. Jean nodded attentively. I moved about looking at the finished pictures that tomorrow or the day after I'd have to deliver to customers.

Jean would never had been able to stand up to their argumentativeness. There was always something they wanted changed more to their own liking. And that was my job—a heavy job certainly—to convince them: The artist is a great French painter, and that alone is enough to guarantee the eternal appeal of the work, to make it more eternal than the customers themselves. If it's changed, that eternal appeal will be destroyed and it will become an ordinary chemical photograph. The most determined querulousness came from the female customers. It's lucky I had been able to learn a lot from listening to Jean: Women prefer to serve the present and are afraid of age; they are in the grip of dreams about their fragile youth and want to hang eternally to the youth of their dreams. Age is truly an oppression for women. You must reply to their arguments in kind: This painting will make an excellent inheritance for Madam's children, not just for Madam. (Luckily our women customers aren't all sterile.) Usually my tenacity wins out. If not, I'm forced to make threats: Well, if Madam doesn't like it, I'll pay it off myself and hang it in my room. Usually this threat arouses their curiosity. They quickly answer:

What for? And I answer: If it becomes my property there are no obstacles to stop me doing anything I like to it. Doing what for example? Yes, well I could give it a mustache . . . (but I've never actually said that). In short, until now I've never been defeated, especially after I realized women look upon argumentativeness as a measure of one's shrewdness.

"It's getting late, Jean, I'm off home."

I climbed over the hedge into the front yard of my boarding house. Darsam had long been waiting for me with a letter.

"Young Master." He offered his respects, then he spoke in Javanese. "Nyai awaits a reply. Darsam will wait, Young Master."

The letter reported that the family at Wonokromo was waiting for me and that Annelies had fallen into daydreaming, did not want to eat, and also that much of her work was unfinished, or incorrectly done. "Sinyo Minke, how grateful would I be, a mother with so much to do, if Sinyo would consider her difficulties. Annelies is my only helper. I cannot handle all the work myself. I'm worried about Annelies's health. It would mean so much to us both. Come and visit, Nyo, even if only for a little while. One or two hours would be enough. Though we hope very much that Sinyo will want to stay with us. Lastly, let me express our unbounded gratitude for Sinyo's cooperation."

The letter was written in correct and proper Dutch. There was no way it could have been written by an inexperienced primary school graduate. I thought it might have been written by somebody else. At least I knew it was not written by Robert Mellema. But who cares who wrote it? I gave me courage, gave me back my character: If I was in their grip, they were also in mine. In each other's grips, if you couldn't actually say under each other's spells. A wise mother, naturally emanating authority like Nyai, is needed by every child. And a maiden whose beauty is beyond compare is needed by every youth. See, I thought: They need me in order to save their family and their business. So I was pretty remarkable too, heh? How many arguments could I now assemble to justify myself in my actions.

Good. I will go.

4

Nyai's letter certainly didn't exaggerate. Annelies looked gaunt. She greeted me on the front steps. Her eyes shone, bringing her face to life again; she had been so pale before she shook my hand.

Robert Mellema wasn't to be seen. And I didn't ask after him.

Nyai emerged from the doorway at the side of the front parlor.

"You've come at last, Nyo. Annelies had to wait so long for you. Look after your friend, Ann; I've still a lot of work, Nyo."

I was able to steal a glance into the room next to the front parlor. It turned out to be nothing other than the business office. Nyai closed the door, and disappeared behind it.

There arose that same feeling I had experienced the first time I came: foreboding. Something strange could happen at any moment. Be careful, this heart of mine reminded me. Be vigilant. As before, now too a voice asked me: Why are you so stupid as to come here? Why don't you go home to your own family if you're tired of boarding? Or find somewhere else to board? Why do you follow the pull of this forbidding house, not resisting, but surrendering unconditionally?

Annelies took me into the room where I had stayed last time. Darsam lifted down my suitcases and bags from the buggy and brought them into the room.

"Let me put your clothes into the wardrobe," said Annelies.

I handed over the suitcase keys and she began to busy herself. She lined up the books on the table; the clothes went into the wardrobe. Then she unpacked my bag. Darsam put the empty suitcase and bag on top of the wardrobe. And Annelies now fixed up the row of books so they looked like a column of soldiers.

"*Mas!*" That was the first time she had called me thus—a call that made my heart pound, making me feel as if I were in the midst of a Javanese family. "Here are three letters. You haven't read them yet. Why don't you read them?"

It felt as if everyone was demanding I read the letters I received.

"Three letters, Mas, all from B———."

"Yes, I'll read them later."

She brought them over to me, saying:

"Read them. They might be important."

She went to open the outside door. And I put the letters on the pillow. I followed after her. In front of us a beautiful garden opened up, not big, you could almost say tiny, with a pool and a few white geese chatting away—like in pictures. A stone bench stood at the edge of the pool.

"Come on." Annelies took me outside, along a cement path hemmed in on either side by the green lawn.

We sat down on the stone bench. Annelies was still holding my hand.

"Does Mas prefer I speak Javanese?"

No, I didn't want to oppress her with a language that would force her to position herself according to the complicated Javanese social order.

"Dutch is fine," I said.

"We had to wait so long for you."

"I had a lot of schoolwork, Ann; I must pass."

"Mas will surely pass."

"Thank you. Next year I must graduate. Ann, I think of you all the time."

She looked at me with a glowing face and pushed her body closer to mine.

"Don't lie," she said.

"Who would ever lie to you?"

"Is it true?"

"Of course. Of course."

I held her around the waist and heard her shallow breathing. Oh Allah, You have given me the most beautiful maiden in the world. My heart raced too.

"Where is Robert?" I asked in an attempt to tranquilize my heart.

"Why ask that? Even Mama never asks where he is."

There was something the matter here, but I didn't feel it was right for me to interfere.

"Mama feels she can't do all her work anymore, Mas." She bowed her head and her voice contained sorrow. "These days I have to carry out all her duties."

I observed her pale, waxen lips.

"Robert doesn't like Mama. He doesn't like me either. He's hardly ever at home. He hates everything Native, except the pleasure he can get from them. It's like he's not Mama's firstborn son, not my brother. He's like a stranger who has wandered off the road into our house."

She obviously thought a lot about her brother, and thought about him with compassion—and she was still so young.

"I haven't seen Mr. Mellema either," I said, looking for something else to talk about.

"Papa? Are you still afraid of him? Forgive that horrible night. You shouldn't think about him anymore. Papa has become such a stranger. Sometimes he only comes home once a week, then leaves again the same day. Sometimes he sleeps a while, then vanishes, I don't know to where. That's why Mama and I have to look after everything now."

What sort of family was this? Two women, mother and daughter, working silently away to maintain a family and a business as big as this?

"Where does Mr. Mellema work?"

"Don't pay any attention to him, I beg you, Mas. No one knows where he works. He never speaks; it's like he's mute. And we never ask. No one talks with him. This has been going on for five years now. It feels that's how it's been for as long as I can remember. He used to be so good and so friendly. Every day he put aside time to play with us. All of a sudden, when I was in fourth class at E.L.S., everything changed. The business closed down for several days. Red-eyed, Mama came to the school to pick me up, to take me away from school forever. Beginning that day I've had to help Mama with her work in the business. Papa never appeared again, except for a few minutes every one or two weeks. Since then, Mama has never really spoken to him, or wanted to answer his questions."

An unhappy story.

"Was Robert also taken out of school?" I asked, turning the conversation.

"When I was taken out of school he was in seventh class—no, he wasn't taken out."

"Where did he continue his schooling afterward?"

"He passed that year, but didn't want to go on. He didn't want to work. Soccer and hunting and horseback riding—that's all he was interested in."

"Why doesn't he help Mama?"

"Because he hates Natives, says Mama. For him there would be nothing greater than to become a European and for all Natives to bow down to him. Mama refuses to bow down. He wants to control the whole business. Everyone would have to work for him, including Mama and me."

"He looks on you as a Native too?" I asked cautiously.

"I *am* a Native, Mas," she answered without hesitation. "You're surprised? Yes, I could call myself an Indo. I love and believe in Mama, Mas, and Mama is a Native."

A puzzling family indeed, each member playing their part in this fearful play.

Annelies kept on talking and I just listened.

"If that's what you want, that's easy, Robert, said Mama. You're an adult now. When your papa dies, go to a lawyer; perhaps you could get control over the whole business. Mama also said: But you must remember, you still have a stepbrother from a legitimate marriage, an engineer called Maurits Mellema, and you would never be able to stand up to a Pure European. You are only

67

a Mixed-Blood. If you really want to own and run this company properly, learn to work like Annelies. You can't even govern workers, because you have never worked yourself."

"Look at the swan, Ann, white, like cotton wool." I wanted to change the conversation. But she kept on talking. "Why are you telling me all your family secrets?"

"Because you are our first guest in five years. Our guest, a family guest. There have been some visitors but they've all been business contacts. There was one man, a family guest, but he was our doctor. So you're our first real guest. And you're so close to us, so good to Mama and also to me." Her voice faded into quietness, no longer childlike. "See, I'm ready to tell you everything, Mas. And you mustn't feel restrained about anything here either. You'll be the good friend of us both." She became very sentimental. "Everything I own is yours, Mas. You are free to do as you wish in this house."

How lonely were the hearts of this girl and her mother in the midst of this abundant wealth.

"Rest now. I want to do some work."

She stood, ready to leave. She looked at me for a moment, hesitated, kissed me on my cheek, then quickly walked away, leaving me by myself, alone.

How long had she been saving up all her feelings? I became the receptacle into which they overflowed.

I could hear the racket from the rice factory from where I was sitting. And the sounds of the milk carts coming and going. The bang and clatter of the buffalo carts as they took things to and from the warehouse. The threatening pounding as peanuts were broken from their shells. The noise of workers joking.

I entered the room, opened up my notebook, and began writing about this strange and frightening family that, by sheer accident, had now involved me too in its affairs. Who knows, I thought, some day in the future I may be able to produce stories like *When the Roses Wilt,* that remarkable serial by Hertog Lamoye? Yes, who knows? So far I've only written advertisements and short articles for the auction papers. With my own byline and read by the public? Who could tell?

I wrote down all of Annelies's words. And what about Darsam, the fighter? I still did not know much about him. With which of this forbidding family's three factions did he side? Wasn't he precisely the closest danger to all three? Danger? Is there really any danger? If there is, then I too must be under threat. If it's true there is danger, why am I staying here? Isn't it better I leave?

The knock on the door startled me. Nyai was standing in front of me.

"We can't tell you how happy we are that Sinyo was prepared to come. See, Nyo, she is beginning to work again, she has got back her liveliness. Sinyo's arrival is not only helping the company, but more importantly, Annelies. She loves you and she needs your attention. Forgive my frankness, Minke."

"Yes, Mama," I answered respectfully, more respectfully, it felt, than to my own mother. And I felt again her black magic gripping me.

"All right, stay here for the time being. I will set aside buggy and driver for Sinyo's use."

"Thank you, Mama."

"So Sinyo is willing to stay here? Why are you quiet? Yes, yes, think it over first. Anyway, Sinyo's here now."

"Yes, Mama," and her hold on me made itself felt even more.

"Good. Rest. Even if a bit late, there's no harm in me congratulating you on doing so well at school."

And so I became a new member of this family. I noted, however, that I must remain vigilant, especially towards Darsam. I will not get too close to him. On the contrary, I must always be polite towards him. There's no doubt that Robert will hate me as a worthless Native. Mr. Herman Mellema will, for sure, spit abuse at me whenever the opportunity arises. In short I must be vigilant—the price to be paid for the happiness of being close to Annelies Mellema. And what can be obtained in this life without payment? Everything must be paid for, or redeemed, even the shortest happiness.

Robert didn't appear at dinnertime. Nor did the shadows and scraping strides of Mr. Mellema.

"Minke, Nyo," Nyai began, "if you like to work and strive, you'll be happy here with us. We also will feel safer with a man around the house. I mean, a man who can be relied upon."

"Thank you, Mama. That is all good and very pleasing, although I still must give it some thought first," and I told them

69

about the situation of Jean Marais's family and how they still needed my services.

"That is good," said Nyai. "It's proper that people have friends, friendships without self-interest. Without friends, life is too lonely." Her words were directed more at herself. Suddenly: "Aha! Ann, Sinyo Minke is now close to you. Look well. He is already close to you. Is there anything else you want?"

"Ah, Mama," Annelies murmured as she glanced across at me.

"Ah Mama! Ah Mama! That's all you can ever say. Come on, speak now, while I can hear too."

Annelies glanced at me again and her face was scarlet. Nyai smiled happily. Then she stared at me, and said:

"Nyo, this one here . . . is like a little child. And what about yourself Nyo, what do you have to say now you're close to Annelies?"

Now it was my turn to be embarrassed into silence. And there was certainly no way I would be calling out "Ah Mama" like Annelies. This woman thought quickly and sharply, able to reach straight into people's hearts, as if it were easy for her to know what lived inside people's breasts. Perhaps there lay the power with which she held people in her grasp, and bewitched them from afar. Let alone from nearby.

"Why are you both silent like kittens caught out in the rain?" She laughed, pleased at her own turn of phrase.

Indeed this was no ordinary nyai. She faced me, an H.B.S. student, without any feeling of inferiority. She had the courage to state her opinion. She was aware of her own strength of character.

We passed away the night listening to Austrian waltzes on the phonograph. Mama read a book. I don't know what book. Annelies sat near me without talking. My own thoughts wandered to May Marais. She'd be happy here, I thought. She liked listening to European music. There was no phonograph at home, nor in my lodgings.

I began to tell them about the little girl who had lost her mother. And the fate of that mother. And the goodness of Jean Marais. And his wisdom. And his simplicity.

Nyai stopped reading, put the book down in her lap, and listened to my story.

I continued my story about Jean Marais. One day he heard that his platoon had received orders to attack a village at Blang Kejeren. They departed early and arrived at the village around nine in the morning. Already, some distance from the village, they had let off shots into the air to frighten away opposition so no battle would be necessary. They fired into the air again while resting under the shade of some trees. A while later they set off again, ready to enter the village. The village was already empty. The platoon entered without coming up against any opposition. No one could be found, not even a baby. They started going into the houses and smashing up whatever they could.

The people had become impoverished during the twenty years of war. There was nothing the soldiers could take as souvenirs. Corporal Telinga ordered that all the houses be burned. Precisely at that moment the Acehnese came into view in the distance, like columns of ants, men and women. They all wore black. They shouted in all sorts of voices, calling to Allah. Then, all of a sudden, a group of young men appeared out of nowhere and attacked Marais's platoon. They ran amok with their daggers. No one knew where they came from. The rifles had no ammunition left. And the black ants in the distance were getting closer. Picking up their wounded, Telinga's platoon rushed from the village. Jean Marais was caught in a bamboo trap. A sharp wooden spike pierced his leg. Telinga was also caught in a trap, but it wasn't so serious. They pulled the wooden spike out of Jean's leg, and he fainted. They ran. No one could be sure what else the Acehnese had planned. A new band of attackers could appear suddenly at any moment. All they could do was run and continue to run. And take the wounded who could travel. Fifteen days later it was clear that Jean Marais was infected with gangrene at the base of his knee. Several months earlier he had lost his lover; now he was to lose a leg, cut off above the knee.

"Bring the child here," said Nyai. "Annelies would like very much to have a little sister. Wouldn't you, Ann? Oh, no, you don't need a little sister, you've got Minke now."

"Ah, Mama," she exclaimed, embarrassed.

I was embarrassed too. That was my last chance. I'd tried to build up myself before this remarkable woman as a man with character, as somebody whole. Every time she spoke, my efforts

71

were brought to nothing. My individuality was being hidden by her shadow. And I knew this could not be allowed to go on forever.

"Mama, permit me to ask." I started my effort to escape from her shadow. "Mama graduated from which school?"

"School?" She tilted her head as if spying on the sky, clearing her memory. "As far as I can remember, I've never gone to school."

"How's that possible? Mama speaks, reads, and maybe also writes Dutch. How is that possible without schooling?"

"Why not? Life can give everything to whoever tries to understand and is willing to receive new knowledge."

Her answer truly startled me. Such words had never been spoken by my teachers.

That night I found it difficult to sleep. My thoughts worked hard trying to understand this woman. Outsiders viewed her as only a nyai, a concubine. Or people respected her only because of her wealth. I saw her from yet another angle: from all that she was capable of doing, from everything she said. I think Jean Marais was right: You must first of all think justly. Don't sit in judgment over others when you don't know the truth of the matter.

There are many outstanding women. But I have met Nyai Ontosoroh. According to Jean Marais's stories, Acehnese women used to descend into the battlefield to fight the Indies Army, ready to die beside their men. So too in Bali. In my own birthplace women peasants worked side by side with the men in the paddy fields. Yet none of these was like Mama—she knew more than just the world of her own village.

And all my school friends knew too that there was another outstanding Native woman, a maiden, a year older than me. She was the daughter of the Bupati of J——, the first Native woman to write in Dutch. Her writing had even been published in literary magazines in Batavia. She was seventeen when her first writings were published, writings in a language other than her mother tongue. Half of my friends denied the accuracy of these reports. How was it possible that a Native, a girl to boot, only an E.L.S. graduate, could write and articulate her opinions in the European manner, let alone have them published in a scholarly magazine? But I believed it and must believe it; it was something that strengthened my belief that I too could do what she had done.

Hadn't I already proved that I could do it? Even if still just initial attempts and in a very small way? It was indeed her example that stimulated me to write.

And now I was coming to know this older woman. She doesn't write articles but she is expert in seizing people in her grasp. She runs a big, European-type firm. She confronts her own eldest son, controls her master, Herman Mellema, trains her youngest child to be a future administrator, Annelies Mellema— the beautiful maiden of whom all men dream.

I would study this strange and frightening family. And one day I'd write.

5

I could not restrain my curiosity to know who this extraordinary Nyai Ontosoroh really was. It was only several months later that I found out from Annelies the story about her mother. After reordering, it came out as follows:

You must surely remember your first visit here, Mas? Who could forget it? Certainly not me. Not in all my life. You trembled as you kissed me in front of Mama. I trembled too. If I hadn't been dragged away by Mama I'd still be standing numb on the steps. Then the carriage dragged you away from me.

Your kiss felt hot on my cheeks. I ran to my room and examined my face in the mirror. Nothing had changed. There had been no chili paste with dinner that night, just a little pepper. Why were my cheeks so hot? I scrubbed and I rubbed. They were still hot. Wherever I sent my eyes off wandering they always ended up colliding with yours.

Had I gone mad? Why was it always you that appeared, Mas? Why did I suddenly feel this loss after you went away?

I changed into my nightclothes and put out the candles, got into bed. But the darkness only made your face clearer. I wanted to hold your hand as I had that afternoon. But your hands were not there. I turned onto my left side, then onto my right, but I still couldn't sleep, hour after hour. I felt there was a pair of hands within my breast whose fingers were tickling me, agitating me. I felt I needed to act. But what? I didn't know. I threw off the blanket and pillow and left the room.

I charged into Mama's room without knocking. As usual, she hadn't gone to bed. She was sitting at her table, reading. She looked at me, closed the book, and I caught a glimpse of the title, *Nyai Dasima*.

"What's the book, Mama?"

She put the object into the cabinet.

"Why aren't you asleep?"

"I want to sleep with Mama tonight."

"A girl old as you, and still wanting to sleep with your mother?"

"Let me, Mama."

"Over there, climb up first!"

I got into the bed. Mama went downstairs to check the doors and windows. Then she came up again, locked the door, pulled down the mosquito net, and put out the candle. Pitch black in the room.

Near her I felt more calm, full of impatient hope, waiting for her words about you, Mas.

"Well, Annelies," she began, "why are you afraid to sleep by yourself? Aren't you grown up by now?"

"Mama, has Mama ever been happy?"

"Even if for only a moment and only a little, everyone has been happy, Ann."

"Are you happy now, Mama?"

"I don't know about now. All there is now is worry. It has nothing to do with the happiness you're asking about. What does it matter if I am happy or not? It's you I worry about. I want to see you happy."

I embraced Mama and I kissed her in the darkness. She was always so good to me. I believed there was no one better.

"Do you love Mama, Ann?"

This question, spoken for the first time, brought tears to my eyes, Mas. She had always seemed so hard.

"Yes, Mama wants to see you happy always. Not ever to feel

75

the pain that I once did. I don't want you to suffer the loneliness I suffer now: without acquaintances, without friends, let alone really close friends. Why all of a sudden do you bring up *happiness*?"

"Don't ask me, Mama, tell me the story."

"Ann, Annelies, perhaps you don't feel it, but I've been deliberately harsh with you so that you would develop the ability to work, so in the future you won't have to be dependent on your husband, if—may it never happen—your husband turns out like your father."

I knew Mama had lost all her respect for Papa. I understood how she felt so I never asked about it. And indeed I wasn't hoping to hear talk about that. I wanted to know if she had ever felt the way I was feeling at that moment.

"When did Mama feel very, very happy?"

"There were many such years after I was with Mr. Mellema, your father."

"Then, Mama?"

"Do you remember when I took you from school? That was when our happiness ended. You're grown up now, it's time you knew. It's best you know what really happened. I've been meaning to tell you for several weeks now. The opportunity has never arisen. Are you sleepy?"

"I'm listening, Mama."

"Your father once said earlier, when you were very, very small: A mother must pass on to her daughter everything she will need to know."

"In those days. . . ."

"Yes, Ann, in those days I respected every word that came from your Papa. I remembered them well. I made them my beacons. Then he changed, and became the opposite of everything he had ever taught me."

"Papa was clever then, Mama?"

"Not only clever, but kind. It was he who taught me everything about farming, business, looking after the livestock, the office work. At first I was taught to speak Malay, then to read and write, then after that, Dutch. Your Papa not only taught me, but then also patiently tested everything he'd taught. He made me speak in Dutch with him. Then he taught me to deal with the

76

bank, lawyers, about trade practices, everything that I've now begun to teach you."

"Why did Papa change so much, Mama?"

"There was a reason, Ann. Something happened. Just that once, then he lost all his goodness, his cleverness, his intelligence, his skills. Broken, Ann, destroyed in an instant. He became someone else, an animal that could no longer recognize his own children or his wife."

Mama didn't continue her story right away. It was as if the tale was an omen about my future, Mas. The world became more and more still. All that could be heard was our own breathing. Probably, if Mama had not been so hard towards Papa—so Mama told me again and again—who knows what might have happened to me. Perhaps something far, far worse than I could ever imagine.

"At the time it happened I thought of taking him to a mental hospital. But I hesitated, Ann. What would people think about you later, Ann? If your father was proven to be mad and was declared by the law as 'under custodial care'? His business, his wealth, and his family would be under the control of an executor appointed by a court of law. Your mama, just a Native, would have no rights over anything, and would not be able to do a thing for her child, you, Ann. All our backbreaking efforts, with never a holiday, would have been in vain. And my giving birth to you, Ann, would have been in vain too, because the law would not acknowledge my motherhood, just because I'm a Native and was not legally married. You understand?"

"Mama!" I whispered. I'd never dreamed the troubles she faced had been so great.

"Even permission for you to marry would not come from me, but from that executor—neither kith nor kin. By taking your papa to a mental hospital, by involving the courts, the condition of your papa would become public knowledge, the public would. . . you, Ann, your fate then, Ann. No!"

"But why would I suffer, Mama?"

"Don't you understand? What would happen to you if everybody knew that you were the daughter of a crazy man? How would we behave in front of everybody?"

I hid my head in the crook of her arm, like a chick.

"His madness was not hereditary," she said to reassure me.

77

"He became so because of a misfortune. But people may not understand that, and you could be thought to have the same lineage." I became frightened. "That's why I've let him be. I know where he's hiding. As long as no one else knows."

Slowly my own problem was pushed aside by my pity for Papa.

"Let me, Mama, let me look after Papa."

"He doesn't know you."

"But he's my papa, Mama."

"Shhh! Pity is only for those who are conscious of their condition. *You* need pity, not him—the child of someone like him. Ann, you must understand: He is no longer a human being. The closer you are to him, the more your life is threatened by ruin. He has become an animal who can no longer tell good from evil. He's no longer capable of any service to his fellow human beings. It's over, don't ask about him again."

I put to sleep my desire to know more. Whenever Mama was serious like that, it wasn't wise to press her further. I wasn't acquainted with any other mothers and children. Both of us had no friends, no comrades. Life as an employer dealing with workers, and as a business person dealing with customers, surrounded by people with no concerns besides business, had left me incapable of making comparisons. I didn't know how other Indos lived. Mama not only didn't allow me to mix sociably, but did not ever leave me the time that would have made it possible. Mama was the only greatness and power that I knew.

"You must understand, don't forget it for as long as you live, the two of us must strive with all our might to make sure no one ever knows that you are the child of a man who has lost his mind." Mama closed the matter.

We were both silent. I didn't know what she was thinking about or imagining. Within my breast those fingers began tickling again. I couldn't stand it. She still hadn't spoken about you, Mas. Did she approve of you or not, Mas? Or were you just to be considered another new factor in the business?

If felt as if the darkness did not exist. Only you existed. Nothing besides you! I had to bring to an end this unpleasant story of Mama's.

"Mama, tell me how you met and then how it was when you lived with Papa."

"All right, Ann. But don't be shocked. You are a spoiled and happy child compared with your mama when she was younger. All right, I'll tell you."

And she began her story:

I had an elder brother, Paiman. He was born on the market day of Paing, so he was named with the first syllable *Pai*. I was three years younger, and named Sanikem. My father changed his name after he was married to Sastrotomo. The neighbors used to say the name meant the foremost scribe.

People said that my father was very industrious. He was respected as the only person in village who could read and write, the sort of reading and writing used in offices. But he wasn't satisfied with just being a clerk in the factory. He dreamed of a higher post, even though the job he held was quite a respected one. He no longer needed to hoe the ground or plow or labor, or plant or harvest sugar cane.

My father had many younger brothers and sisters as well as cousins. As a clerk he had great difficulty in getting them jobs at the factory. A higher post would have made it easier, and also it would have raised him up higher in the eyes of the world, especially as he wanted his relations to be able to work in the factory as something more than just laborers and coolies. At the very least they should be foremen. You didn't need a blood relative as a clerk to get jobs as coolies—anybody could get a job as a coolie as long as the foreman agreed.

He worked diligently and became even more diligent for more than ten years. But still no promotion, though his salary and commission rose every year. So he tried every other way: the traditional Javanese magic men, the *dukuns*; magic formulas; he even went on rice fasts, Monday and Thursday fasts. Still no result.

He dreamed of becoming paymaster: cashier, holder of the cash of the Tulangan sugar factory in Sidoardjo. And who did not have business with the factory paymaster? There were the cane foremen: They came to receive their money and leave their thumbprints. If the foreman refused to accept a toll on the coolies' wages, he could withhold the foreman's gang's weekly wages. As paymaster he would be a big man in Tulangan. Merchants would bow down in respect. The Pure and Mixed-Blood tuans would greet him in Malay. The stroke of his pen meant money! He would be counted among the powerful in the factory. People would listen to his words—"Sit down on the bench there"—in order to receive their money from his hands.

Pathetic. These dreams did not bring him a rise in position, respect, or esteem. On the contrary, they brought hatred and disgust. And the position of paymaster remained hanging in limbo, far away. His crawling behavior, which often harmed his friends, caused him to be cut off from society. He was isolated in the midst of his own world. But he didn't care. He was indeed hard-hearted. His trust in the generosity and protection of the white-skinned tuans could not be broken. People were sickened to see the things he did to get the Dutch tuans to come to his house. One or two did turn up and he served them with everything that pleased them.

But the post of paymaster still did not come his way.

He even went as far as using a dukun magic man and ascetic practices to cast a spell on the tuan administrator, the Tuan Besar Kuasa, the "Great, Powerful Tuan," to come to the house. Also to no avail. On the other hand, he often visited the Tuan's house, not to see the official on some business but to help with the manual work in the back of the house! The tuan administrator never took any notice.

I was revolted by all this. Sometimes I would watch my father and feel moved. His whole body and soul wrestled with that dream. How he humiliated himself and his dignity! But I didn't dare say anything. Sometimes I did pray that he would stop his shameful behavior. The neighbors often said it was better, and indeed best, to ask of Allah, for how great anyway was mankind's power; but it was undignified to beg from the white people. I did not pray that he obtain his post, but rather that he be able to shake himself free from his shameful behavior. At that time I would not have been able to explain all this. It was something I just felt inside me. But all my prayers were to no avail.

Tuan Besar Kuasa was a bachelor, as was usually the case with newly arrived Pures. He was, perhaps, older than my father, Sastrotomo, the clerk. People said my father once tried to offer him a woman. He not only rejected the offer and refused to say thank you but abused my father and threatened to sack him. After that my father became the object of public ridicule. My mother quickly grew thin and frail as she listened to people's taunts: "Maybe he'll end up offering his own daughter." They meant me.

You must certainly be able to imagine how suffocating life became after that. From that time on I never dared leave the house. My eyes were always looking uncontrollably to the front room to see if there was a white-skinned guest. Thanks be to God, there never was.

Unlike the other Dutch men, Tuan Besar Kuasa didn't like participating in the tayub dance festivities. Every Sunday he went to Sidoardjo for devotions at the Protestant church. At seven in the morning he could be seen on a horse or in a carriage. I myself once saw him from afar.

When I turned thirteen I was kept at home, and was only acquainted with the kitchen, back parlor, and my own room. All my friends had already married. Only when a neighbor or relative visited us did I ever feel that I had the freedom of the house as I had in my childhood. I wasn't even allowed to sit on the porch. Not even to step onto its floor.

When the factory stopped work and the employees and workers went home, I often watched from inside as someone would walk back and forth, all the time glancing at our house. Of course, all our lady guests said I was beautiful, the flower of Tulangan, the blossom of Sidoardjo. And if I looked at myself in the mirror, I found no reason not to agree with their flattery. My father was a handsome man. My mother—I never knew her name—was a pretty woman and knew how to look after her body. Actually my father should properly have had two or three wives, especially as he owned land that was rented by the factory and other land worked by tenants. But he didn't. He felt it was enough to have one wife who was beautiful. His only other dream was to become the paymaster, factory cashier, the most respected Native, for the rest of his life.

That was how things were, Ann.

By the time I reached fourteen, people already considered me an old maid. I'd already begun having periods two years earlier. Father had some special plan for me. Even though people hated him, proposals of marriage to me came in often. All were refused. From my room I heard it happen several times. Unlike other

81

Native women, my mother had no say in any of this. Father decided everything. My mother did once ask what sort of son-in-law he was hoping for. He didn't answer.

No, Ann, I'm not going to be like my father and decide what sort of son-in-law I must have. You choose; I only advise. But that was my situation, Ann, the situation of all young girls then—they could do nothing else but wait for a man to take them from the house, to who knows where, as wife number who knows what, first or fourth. My father and my father alone determined everything. You were lucky indeed if you turned out to be the first and only wife. And that was an extraordinary thing in a factory area. There was more. The girl never knew beforehand whether the man would be young or old. And once married, the girl had to serve this man, whom she had never met before, with all her body and soul, all her life, until she died or until he became bored and got rid of her. There was no other way, no choice. He could be a criminal, a drunkard, and gambler. No girl would know until after she became his wife. You were lucky if the one who came was a good man.

One night, the tuan administrator, Tuan Besar Kuasa, came to the house. I was on edge. My father was rushing here and there giving orders to Mother and me to do this and that, and then canceling them with still other orders. He ordered me to put on my best clothes and once or twice came in to watch me putting on my makeup. I was indeed suspicious—maybe the people's whisperings were true. My mother was even more suspicious. Nothing had happened yet, but she was already crying and sobbing in a corner of the kitchen, silent in a thousand tongues.

My father, Clerk Sastrotomo, ordered me to come out and serve strong coffee and milk, and cakes. My father had, of course, already given orders: Make the coffee strong.

I came out carrying a tray. The coffee and cakes were on it. I didn't know what Tuan Besar Kuasa's face was like. It was not proper for a well-mannered girl to lift up her eyes and face towards a male guest who was not known well to the family, especially if he was white. I kept my head down, placing the contents of the tray on the table. Even so, his trousers were visible; they were made from white drill cloth. And his shoes: big, long. A sign that the man was tall and big.

I felt the eyes of Tuan Besar Kuasa pierce my hands and my neck.

"This is my daughter, Tuan Besar Kuasa," my father said in Malay.

"It's time she had in-laws," responded the guest. His voice was big, heavy and deep, as if it came from his whole chest. No Javanese had a voice like that.

I withdrew again to await new orders. And no orders came. And then the Tuan Besar Kuasa left with father, who knows to where.

Three days later, after lunch, midday Sunday, Father called me. He sat with Mother in the central room. I knelt in front of him.

"Don't, Papa, don't," Mother protested.

"Sanikem, Kem, Ikem," Father began, "put all your possessions and clothes into your mother's suitcase. Dress yourself well, neatly, attractively."

Ah, how many questions attacked my heart! I must carry out all my parents' orders, especially Father's. I could hear my mother protest and protest, but father took no notice. I packed all my clothes and possessions. My clothes could have been considered expensive and numerous, compared to those of other girls, so I had looked after them well. I had more than six batiks. Among them were some I had made myself.

And I came out carrying the old brown suitcase, with its dents here and there. Father and Mother were still sitting in the same place. Mother refused to change clothes. Then all three of us left in the carriage that was waiting in front of the house.

Once in the carriage, Father spoke, his voice clear and free of hesitation.

"Look at your home, Ikem. From today this is no longer your home."

I had to be able to understand his meaning. I heard Mother sobbing. I was indeed being expelled from my home. I also wept.

The carriage stopped in front of Tuan Besar Kuasa's house. We all got down. That was the first time Father did anything for me: He carried my suitcase.

I didn't dare look around me. Yet I felt there were thousands of pairs of eyes staring at us in amazement.

I just stood there at the top of the steps of that stone house.

83

My thoughts and feelings only added to my burden, sucking everything from my body. All that was left of my body was its skin. So in the end I was being brought here. Truly, Ann, I was ashamed to have as a father Sastrotomo the clerk. He was not fit to be my father. But I was still his daughter, and there was nothing I could do. Neither the tears nor the tongue of my mother could prevent the disaster. Let alone I, who neither understood nor owned this world. I did not even possess my own body.

Tuan Besar Kuasa came out. He smiled happily and his eyes were bright. And I heard his voice. In a foreign sign-language he invited us to come up. In a flash it became clearer to me just how big and tall his body was. Perhaps three times as heavy as Father. His face was reddish. His nose protruded very much, enough at once for three or four Javanese noses. The skin on his arms was coarse like an iguana's skin, and was thick with yellow hair. I gnashed my teeth, bowed my head down further. His arms were as big as my legs.

So it was true that I was to be surrendered to this white, iguana-skinned giant. I must be strong, I whispered to myself. No one is going to help you! All the devils and demons had encircled me.

For the first time in my life, at the invitation of Tuan Besar Kuasa, I sat on a chair the same height as Father. Before the three of us: Tuan Besar Kuasa. He spoke in Malay. I could catch only a few words. During the conversation everything felt as if it were surging and then falling again like the ocean. I could not find a moment of peace. Tuan Besar Kuasa took out an envelope from his pocket and handed it to Father. He also took out a piece of paper with writing on it and Father put his signature to it. Afterwards I found out the envelope contained twenty-five guilders, representing Father's surrender of me to him, along with the promise that Father would be made cashier after first successfully completing a two-year trial period.

So, Ann, that was the simple ritual whereby a child was sold by her own father, Clerk Sastrotomo. And who was it who was sold: I, myself, Sanikem. From that moment on I lost all respect and esteem for my father—for anyone who has ever sold their own children, for whatever purpose or reason.

I kept my head bowed down, knowing that no one would be able to take up my cause. In this world only Father and Mother held power. If Father was as he was, if Mother could not defend me, what could anyone else do?

Father's final words:

"Ikem, you must not leave this house without the permission of Tuan Besar Kuasa. You may not return home without his permission and without my permission."

I did not look at his face as he spoke those words. I kept my head bowed.

Father and Mother went home in the same carriage. I was left on the chair, bathing in my own tears, shaking and not knowing what I must do. The world seemed dark. Looking up from under my bowed head, my vision blurred, I could still see Tuan Besar Kuasa as he entered the house after having said farewell to my parents. He picked up my suitcase and took it into a room. He came out of the room and approached me. He pulled my hand, ordering me to stand. I trembled. It wasn't that I didn't want to stand up, or that I was rebelling against an order. I didn't have the strength to stand. My kain was soaking with sweat. My legs trembled so badly, it was as if my bones and sinews had come loose from their joints. He picked me up as if I were an old pillow, carried me in his arms into the room, and put me down on a beautiful, clean bed, powerless. I was not able even to sit. I rolled over; perhaps I fainted. But my eyes could still vaguely make out the room. Tuan Besar Kuasa opened my suitcase and put my clothes into a big wardrobe. He wiped the suitcase with a cloth and put it inside the bottom section.

He returned to me, where I had rolled over prostrate on the bed.

"Don't be afraid," he said in Malay. His voice was low like thunder. His breath blew in my face.

I closed my eyes tightly. What was this giant going to do to me? He picked me up and carried me around the room like a wooden doll. He took no notice of my wet kain. His lips touched my cheeks and my lips. I could hear his breath, which blew hard into my ears. I dared not cry. I dared not move. My whole body was soaked in a cold sweat.

He stood me on the tiles. He caught me again when he saw that I slumped, about to collapse. He picked me up again and hugged me and kissed me. I can still remember his words, though I didn't then understand them:

"Darling, my darling, my doll, darling, darling."

He threw me up and caught me on his hip. He rocked me, and that way I got back some of my strength. He stood me up again on the floor. I swayed and he guarded me with his hand so I would not fall. I still swayed, falling headlong onto the edge of the bed.

He strode towards me, and opened my lips with his fingers. With signs he ordered me to brush my teeth. He walked me outside to the back of the house, to the bathroom. That was the first time I saw a toothbrush and how to use it. He waited until I finished, and my gums hurt all over.

Again, with signs, he ordered me to bathe and to scrub myself with scented soap. I carried out all his orders as if they were from my own parents. He waited for me outside the bathroom with sandals in his hands. He put the sandals on my feet. Very, very big—the first sandals I ever wore in my life—made from leather, heavy.

He carried me inside the house to the room and sat me in front of a mirror. He rubbed my hair with a thick cloth until it was dry. I later found out the cloth was called a towel. Then he put oil on my hair—its scent was so fragrant. I didn't know what kind of oil it was. And it was he too who combed my hair, as if I couldn't. He tried to set my hair in a bun, but he couldn't and he let me finish it.

Then he instructed me to change clothes, and he observed all my movements. I felt I had no soul anymore, like a shadow puppet in the hands of the puppet master. After I had finished dressing, he powdered my face. Then he put a little lipstick on my lips. He took me out of the room and called his two maidservants.

"Look after my nyai well!"

And such was my first day as a nyai, a concubine, Ann. And it turned out that his caring and friendly actions drove away some of my fears.

After giving the orders to his servants, Tuan Besar Kuasa left. Who knows to where. The two women babbled about how lucky I was to be taken as a nyai. I did not want to say anything. I didn't know this house and its customs. Of course there was in my heart the wish to run. But from whom could I seek protection? Then what would I do? I didn't dare. I was in the hands of someone very powerful, more powerful than Father, than all the Natives in Tu-langan.

They prepared food and drinks for me. Every other minute they knocked on the door, offering this and suggesting I take that. I was mute, just sitting on the floor, not daring to touch anything in the room. My eyes were open, but I was afraid to look: Perhaps this was death in life.

That night Tuan came. I heard the steps of his shoes as they came nearer. He came straight into the room. I shuddered. The lamp, which the servants had lit earlier in the evening, threw the light onto his clothes, all white and dazzling. He came up to me. He picked up my body from the floor, put it on the bed, and laid it down there. It seemed I dared not even breathe, afraid that I might enrage him.

I don't know how long that mountain of flesh was with me. I fainted, Annelies. I didn't know any longer what was happening.

As soon as I regained consciousness, I knew I was no longer the Sanikem of the previous day. I'd become a real nyai. Later I found out the name of that Tuan Besar Kuasa: Herman Mellema. Your papa, Ann, your true papa. And the name Sanikem disappeared forever.

You're already asleep? Not yet? Not yet?

Why am I telling you this story, Ann? Because I don't want to see my child go through such cursed experiences as these. You must marry properly. Marry someone like you, of your own will. You, my child, you must not be treated like a piece of livestock. My child may not be sold to anyone, no matter what the price. Mama will make sure that such a thing does not happen to you. I will fight to preserve the dignity of my child. My mother was incapable of defending me, so she was not fit to be my mother. My father sold me like the offspring of a horse; he wasn't fit to be my father. I don't have any parents.

Life as a nyai is very, very difficult. A nyai is just a bought slave, whose only duty is to satisfy her master. In everything! Then, on the other hand, she has to be ready at any moment for the possibility that her master, her tuan, will become bored with her. And she may be kicked out with all her children, her own children, unrecognized by Native society because they were born outside wedlock.

I swore in my heart I would never look upon my home or my parents again. I did not even care to remember them. I never wanted to think about that humiliating event again. They had made me into a nyai like this. So I must become a nyai, a bought slave, a good nyai, the very best nyai. I studied everything possible about my master's wants: cleanliness, Malay, making the bed, ordering the house, cooking European food. Yes, Ann, I would have revenge upon my parents. I had to prove to them that whatever they had done to me, I would be more worthy of respect than they, even if only as a nyai.

Ann, I lived for one year in the house of Tuan Besar Kuasa, Herman Mellema. I never went out, was never taken anywhere or met any guests. What would have been the point anyway? I myself was ashamed to meet the outside world. Especially acquaintances, neighbors. I was even ashamed to have parents. I ordered all the servants to go. I did all the housework myself. There were to be no witnesses to my life as a nyai. There were to be no reports about me: a degraded woman, without value, no real will of her own.

Clerk Sastrotomo came to visit several times. I refused to receive him. Once his wife came. I wasn't even prepared to look at her. Mr. Mellema never chided me for my behavior. On the contrary, he was very satisfied with everything I did. It appeared he was very pleased with how I liked to learn. Ann, your papa cared for me very much. But all that could not mend my wounded pride and self-respect. Your papa remained a stranger to me. And indeed your Mama never made herself dependent on him. I always looked upon him as someone I did not really know, who could leave home for the Netherlands at any moment, leaving me, and forgetting everything about Tulangan. Indeed, I prepared myself precisely for that eventuality. If Tuan Besar Kuasa went away, I had to be able not to return to the house of Sastrotomo. Mama learned to be thrifty, Ann, to save. Your papa never asked how I used the shopping money. He himself used to buy the monthly provisions in Sidoardjo or Surabaya.

Within a year I had saved more than a hundred guilders. If one day Mr. Mellema went home, or got rid of me, I'd have capital to take to Surabaya and would be able to begin to trade in whatever I liked.

After I'd lived with Mr. Mellema a year, his contract expired.

He didn't extend it. Ever since he arrived in Tulangan, he had kept dairy cattle from Australia and I was taught how to look after them. In the evening I was taught to read and write, to speak and to put together Dutch sentences.

We moved to Surabaya. Mr. Mellema bought a large piece of land in Wonokromo, our place here now, Ann. But it wasn't as busy as this then, still bush and patches of new jungle. The cattle were moved here.

At that time I began to feel glad, happy. He always paid attention to me, asked my opinion, invited me to discuss everything with him. Gradually I came to feel I was equal to him. I was no longer ashamed if I had to meet with old acquaintances. Everything that I had learned and done during that year had restored my self-respect to me. But my outlook was still the same: I readied myself to be no longer dependent on anyone. Of course, it was going too far for a Javanese woman to speak about self-respect, especially one as young as I was then. It was your papa who taught me, Ann. It was much later that I was able truly to feel the meaning of that self-respect.

Father also visited us at our new address several times, but I still refused to meet him.

"Meet your father," Mr. Mellema ordered. "No matter what, he's still your father."

"I did indeed have a father, once, not any more. If he wasn't Tuan's guest, I would have already thrown him out. It'd be better to leave here than meet him."

"If you went, what would happen to me? What about the cattle? There'd be nobody to look after them."

"There are many people who you can hire to look after them."

"Those cows only know you."

So it was that I began to understand that in reality I was not at all dependent on Mr. Mellema. On the contrary, he was dependent on me. I then began to take a role in making decisions on all matters. He never rejected this. He never forced me to do anything, except study. In this matter, he was a hard but good teacher. I was an obedient and good pupil. I knew everything he was teaching me would, one day, be of use to me and my children if he went home to the Netherlands.

Tuan never pressured me about Sastrotomo again. Several

times that clerk passed on messages through Tuan that if I was unwilling to meet him, then could I write him a letter? I never responded to any of this. I never wrote even one or two lines to him, even though I could now write well in both Malay and Dutch. Sastrotomo wrote again and again. I never read any of his letters; I just sent them back.

Then, once, Mother and Father came to Wonokromo. Tuan was uneasy, perhaps embarrassed, because I still refused to meet them. The guests, according to Tuan, had kept pressing to meet me. Mother had cried. Through Tuan, I said:

"Consider me your egg that has fallen from the egg rack. Broken. It's not the egg's fault."

With that, all business between my parents and me was over.

Why are you squeezing my arm, Ann? I've brought you up to be a business woman and merchant. You should not be sentimental. Our world is one of profit and loss. You don't agree with your Mama's attitude, do you? Even fowls, and the hen most of all, of course, defend their chicks, even against the eagle in the sky. It was only proper that my parents received fitting punishment. You too may take that attitude towards Mama one day. But later, when you're capable of standing on your own feet.

Tuan then imported more cattle, also from Australia. The work increased. Workers had to be hired. Tuan began to hand over all the work within the business to me. At first I was afraid to give the workers orders. Tuan guided me. He said: Their employer is their livelihood. You are the master of their livelihood! Then, under his supervision, I began to dare to give orders. He remained a hard and wise teacher. No, he never hit me. Had he done so just once, my bones would have shattered. As difficult as it was, I was slowly able to do what he wished.

Tuan himself spent most of his time away looking for customers. Our business began to flourish.

At that time Darsam arrived, an unemployed vagrant. But he loved to work. He would do anything that was given to him. One night, after a knife fight, he caught a thief. The thief died. Yes, there was a court case, but he was freed. Since then I have put my trust in him. I've made him my right-hand man. In the meantime Tuan spent less and less time at home.

I almost forgot to tell you this too, Ann. It was Tuan who taught me how to dress properly, to choose matching colors. He liked to wait upon me while I put on my makeup. On those occasions he would say:

"You must always be beautiful, Nyai. A crumpled face and untidy clothes also reflect a crumpled and untidy business. No one will trust you."

See how I fulfilled all his desires? How I satisfied all his needs? I was always neat and tidy. Sometimes I even put on makeup before sleeping. To be attractive and beautiful is truly better than being crumpled, Ann. Remember that. And nothing bad is attractive. If I were a man, I'd tell my friends that any woman who was unable to look after her own beauty was not worth marrying: "She can't do anything, she can't even look after her own skin," I would say.

Tuan said:

"You are not allowed to chew betel nut, that way your teeth will stay gleaming white."

And I never chewed betel nut.

Ann, almost every month books and magazines arrived from the Netherlands. Tuan liked to read. I don't know why you're not like your father, especially when I like to read too. None of this reading material was in Malay, let alone Javanese. When the work was finished, at twilight, we'd sit in front of our hut, a bamboo hut, Ann—we didn't have this beautiful house then—and he would order me to read. Also newspapers. He listened to my reading, correcting me when I was wrong, explaining the meaning of words I didn't understand. So it went every day until eventually I was taught how to use a dictionary by myself. I was only a bought slave. I had to do everything as he wished. Every day. Then he gave me a quota of reading. Books, Ann. I had to finish them and retell their contents.

Yes, Ann, as time went on, the old Sanikem began to disappear completely. Mama grew up into a new person with a new vision and new views. I no longer felt like the slave who was sold years before in Tulangan. I felt as if I no longer had a past. Sometimes I asked myself: Had I become a Dutch woman with brown skin? I didn't dare answer, even though I saw the backwardness of the Natives around me. I didn't mix very much with Europeans, except with your papa.

I once asked him if European women were taught as I was now being taught. Do you know how he answered?

"You are far more capable than the average European woman, especially the Mixed-Bloods."

Ah, how happy I was with him, Ann! How clever he was at flattering me and encouraging me. I was ready to surrender my whole body and soul to him. If my life was to be short, I wanted to die in his arms, Ann. I knew my decision to cut off my links with the past was right. He fitted exactly the Javanese description of a husband: instructor and god. Perhaps to prove what he said, he subscribed to some women's magazines from the Netherlands for me.

Then Robert was born. Four years later, you, Ann. The business got bigger. Our land grew larger. We were able to buy some wild forest at the edge of our land. All the land was bought in my name. There were no rice paddies or other fields yet. After the business became very large, Tuan began to pay me for my labor, as well as for the years that had already gone by. With that money I bought a rice mill and other plants and equipment. Since then the business was no longer the property of Mr. Mellema as my master, but also my property. Then I received a share of five years' profit, five thousand guilders. Tuan obliged me to save it in a bank under my own name. By then we had named the business Boerderij Buitenzorg. And because I carried out all its affairs, people who had dealings with me called me Nyai Ontosoroh, Nyai Buitenzorg.

Asleep yet? Not yet? Good.

After following the women's magazines for a long time and carrying out much of what they taught, I repeated my question to Tuan:

"Am I like a Dutch woman yet?"

Your papa laughed broadly.

"It's impossible for you to be like a Dutch woman. And it's not necessary either. It's enough that you are as you are now. Even thus, you're cleverer and better than all of them. All of them!" He laughed expansively again.

Of course he was exaggerating. But I was pleased and happy. At least I was not below them. It pleased me to hear his praises. He never criticized, there was nothing but praise. He never ignored my questions; they were always answered. I became more confident, more daring.

Then Ann, then this happiness was horribly shaken, rocking

92

the foundations of my life. One day Tuan and I went to court to acknowledge Robert and you as the children of Mr. Mellema. In the beginning I thought that with such acknowledgment my children would receive legal recognition as legitimate children. But it wasn't so, Ann. Your elder brother and you continued to be considered illegitimate, but now you were recognized as the children of Mr. Mellema and could use his name. However, the court's decision also meant that the law no longer recognized you as my children. You weren't my children any longer, though it was I who gave birth to you. Since that day, both of you, according to the law, were the children of Mr. Mellema alone—according to the law, Ann, Dutch law in these Indies. Don't be mistaken. You're still my child. Only then did I realize how evil the law was. You obtained a father, but lost a mother.

Following that, Ann, Tuan wanted you both to be baptized. I didn't go with you to church. You all returned home quickly. The priest refused to baptize you. Your papa became gloomy.

"These children have the right to a father," said Tuan. "Why don't they have the right to receive the absolution of Christ?"

I didn't understand such matters, and was silent. Afterwards I found out that you could only become legitimate if we married in a civil registry office. You could then be baptized, so I began to press your papa that we marry. I urged and urged. Your papa, who had been depressed over those last few days, all of a sudden became angry, the first time he'd ever been angry in all those years. He didn't answer. And he never explained the reasons either. So, according to the law, you are still illegitimate children. And you've never been baptized either.

I didn't try again, Ann. I had to be happy with things as they were. No one would ever call me Mrs. The title Nyai would follow me forever, for all my life. It didn't matter as long as you both had a respected father, who could be relied upon, who could be trusted, who had honor, especially as that acknowledgment meant a great deal in your own society. My own interests need not concern anyone, as long as you both obtained what was rightfully yours. My interests? I could look after them myself. Ah, you're asleep.

"No, Mama," I denied.

I still awaited her words about you, Mas. At another time, when another opportunity arose, perhaps she would not talk as much as this. So I had to be patient until she turned to talking about our relationship, Mas.

So I asked, leading her on:

"So Mama eventually loved Papa too."

"I don't know what the meaning of love is. He carried out his responsibilities, so too did I. That was enough for us both. If, in the end, he had gone home to the Netherlands, I would not have tried to stop him, not only because I had no right to do so, but because we owed each other nothing. He could leave any time he liked. I felt strong with everything I'd learned and I'd obtained, everything I owned and could do. Anyway, Mama was just a concubine whom he'd bought once from my parents. My savings amounted to more than ten thousand guilders, Ann."

"Mama never visited the family in Tulangan?"

"I had no family in Tulangan. Only in Wonokromo. My brother Paiman visited me several times and I would receive him. He came to ask for help. It was always the same. The last time he came was to report: Sastrotomo died in a cholera epidemic along with all the others. His wife had died earlier, who knows of what."

"Perhaps it would be better if we visited there, Mama?"

"No, it's better the way it is now. Let the past be severed from the present. The wounds to my pride and self-respect still haven't healed. If I remember how I was so humiliatingly sold . . . I'm not able to forgive the greed of Sastrotomo and the weakness of his wife. Once in their lives people must take a stand. If not, they will never become anything."

"You're too hard, Mama, too hard."

"And what would become of you if I wasn't ready to be hard? Hard towards everybody. In this, let me be the only victim; I've already accepted my fate as a slave. You're the one who is too weak, Ann, showing pity where it's out of place."

And Mama still hadn't talked about you. It seemed that Mama had never loved Papa, so I was embarrassed to talk about it, Mas. Papa remained a stranger to Mama. While you, Mas, why were you so close to me now? And why did I always want to be near you?

"Then the second blow fell, Ann," Mama continued. "And the wound was never to heal."

*   *   *

The government decided to repair and upgrade Tanjung Perak, Surabaya's harbor. A team of harbor engineers was brought out from the Netherlands. At that time our dairy business was expanding well. Every month there were more and more requests for regular deliveries. The whole D.P.M., the Dutch oil company, was ordering from us. All of a sudden, like a thunderbolt, disaster crashed down. A thunderbolt of disaster.

(Mama went downstairs to get something to drink. The room was dark. No one was listening to us in that upstairs room. It was a still night. The tick-tock of the pendulum clock could be faintly heard coming from the living room through the open door. And that sound disappeared when Mama came back and closed the door.)

There was a young engineer in that team of experts. I first read his name in the newspaper: Engineer Maurits Mellema. A little of his life story was presented. He was a hard-headed engineer. Already in his short career he had proved his great ability, it said.

Perhaps he's your papa's family, I thought. I didn't want anybody else mixing in our lives, which had become so calm, stable, and happy. Our business must not be touched by anyone. So I hid the paper before your papa was able to read it. I said it hadn't arrived—perhaps the delivery man was sick. He didn't ask further about it.

Three months later, Mama went on, after you and Robert had left for school, a guest arrived in a big, beautiful government carriage pulled by two horses. Your papa was working out back. Mama was working in the office.

The government carriage stopped at the front steps. I left the office to greet it. Perhaps some government office needed dairy products. I saw a young European alight. He was dressed all in white. His coat was white, closed, the coat of a marine officer. He wore a marine cap, but there were no marks of rank on his sleeves or shoulders. His body was straight and his chest broad. He unhesitatingly knocked on the door several times. His face was identical to Mr. Mellema's. The silver buttons on his shirt gleamed with pictures of anchors.

In bad Malay, he spoke abruptly and arrogantly, in a manner I felt straight away to be impudent and opposed to the European politeness I knew.

"Where's Tuan Mellema," he said, more an order than a question.

"And you are Tuan who?" I asked, offended.

"I only need to meet Tuan Mellema," he said more roughly than before.

I felt again like a nyai without the right to be respected in my own home. As if I weren't a shareholder in this big business. Perhaps he thought I was sponging on Mr. Mellema. Without my help my master would never have been able to build this house, Ann. This guest did not have the right to act so arrogantly.

I did not invite him to sit down and left him standing. I ordered somebody to fetch Tuan.

Your father taught me never to read a letter or listen to a conversation with which one had no business. But this once I was suspicious. I left the door between the office and the front room open a little. I had to know who he was and what it was he wanted.

The young man was still standing when Tuan came. Through the gap left by the open door I saw your papa stand nailed to the floor.

"Maurits!" Tuan greeted him. "You're already so dashing."

At once I knew this was Engineer Maurits Mellema, the member of the team of harbor construction experts at Tanjung Perak.

He didn't answer respectfully, Ann, but corrected your Papa arrogantly.

"En-gin-eer Maurits Mellema, Mr. Mellema!"

Your papa looked taken aback at receiving the correction. The guest still stood there. Your papa invited him to sit down, but he didn't respond or sit.

You must listen to this story well, Ann, you must not forget it. Not only because your grandchildren must know, but because his arrival was the source of all your and my difficulties today. And the business's.

The young Dutch guest said:

"I didn't come here to sit down in this chair. There is something more important than sitting. Listen, Mr. Mellema. My mother, Mrs. Amelia Mellema-Hammers, after you left in such a cowardly manner, had to work, breaking her back to sustain me, to educate me, until I graduated as an engineer. I and Mrs.

Mellema-Hammers had resolved no longer to hope for your return, Mr. Mellema. As far as we were concerned, you had disappeared, swallowed up by the earth. We sought no reports of your whereabouts."

Through the gap in the door, the side of your papa's face was visible. He raised his hands. His lips moved but no voice came out. His cheeks trembled uncontrollably. Then his hands fell.

Ann, Engineer Mellema spoke like this:

"You, sir, left behind the accusation that Mrs. Amelia Mellema-Hammers had been unfaithful. I, her son, share her feelings of humiliation. You never brought this matter to court. You never gave my mother the opportunity to defend herself and her honor. Who knows to whom else you have passed on or told your dirty accusations. By coincidence I'm now serving in Surabaya, Mr. Mellema. By coincidence also I read in an auction paper an advertisement offering dairy goods and milk produced by Boerderij Buitenzorg with your name displayed. I hired a detective to find out who you were. Yes, H. Mellema was Herman Mellema, the husband of my mother. Mrs. Amelia Mellema-Hammers could have married again and lived happily. But you, sir, left the matter hanging."

"She could have gone to the court any time she liked if she wanted a divorce," answered your papa very weakly, frightened of his own son, who'd become so wild.

"Why should it be Mrs. Mellema-Hammers when it was you who made the accusations? If you really believe my mother was unfaithful, why don't you, sir, file for divorce right now."

"If it had been me that took the case to court, your mother would have lost all her rights over my dairy business in Holland."

"Don't give us all these ifs and buts, Mr. Mellema. The fact is that you never took the case to court. Mrs. Mellema-Hammers became the victim of your ifs and buts."

"If your mother had not objected to opening up the scandal to the public, I would have done something long ago, and without your advice."

"In those days my mother could not afford to hire lawyers. Now her son is ready and able; yes, even to hire the most expensive. You can open the case. You're also rich enough to hire them, and wealthy enough to pay alimony."

Ann, it was now clear. Engineer Mellema was none other

than your papa's only legitimate son from his legitimate wife. He came as a destroyer to ruin our lives.

I trembled as I heard all this. Clerk Sastrotomo and his wife were never allowed to disturb my children's lives; neither was Paiman. Nor would a change in attitude by Mr. Mellema, should he ever change—or even a change by any one of my children. The family and business had to remain as they were. Now there came your stepbrother, who didn't want just to disturb our lives. He came attacking, aiming to ruin everything.

Up until that time I hadn't said anything. Unable to bear hearing what was being said, I came out to cool down the atmosphere. Naturally I had to help Tuan.

"The detective gave me a very detailed and trustworthy account of things," he continued, ignoring my presence. "I know what is in every room of this house, how many workers you have, how many cattle, how many hectares of paddy and how many tons of other crops you get from your fields, how much is your annual income, how much is in your bank account. And the most fantastic thing in all this concerns the foundation of your way of life, Mr. Mellema, you who accused Mrs. Amelia Mellema-Hammers of unfaithfulness. What's the reality now? Legally, sir, you are still the husband of my mother. But you have gone ahead and taken a Native woman as your bedmate, not for one or two days, but for years! Night and day. Without a legitimate marriage. You, sir, have been responsible for the birth of two *bastard* children!"

On hearing that, my blood rose to my head. My lips trembled, dry. My teeth ground together. I stepped slowly towards him and was ready to claw at his face. He had insulted all that I had secured, looked after, striven after, and had loved all this time.

"Words that are only fit to be heard in the house of Mellema-Hammers and her son!" I retorted in Dutch.

He wasn't even prepared to look at me, Ann. To him I was no more than a piece of firewood. As far as he was concerned, I had committed sin with his father and his father had committed sin with me. Perhaps it was his right and the right of the whole world to make such a judgment. But that your papa and I had cheated on this woman Amelia, whom I'd never known, and her child . . . that was the height of his vulgar impudence. And it was happening in the house we ourselves had worked for and built, in our own home.

"You have no right to talk about my family," I roared in Dutch.

"I have no business with you, Nyai," he answered in Malay, pronounced very coarsely and stiffly. He refused to look at me again.

"This is my house. You can speak like that out on the street, not here."

I signaled to your father that Maurits should go, but he didn't understand. Meanwhile this impudent man continued to ignore me. Your papa just stood there open-mouthed, like somebody who'd lost his senses. And it turned out later that indeed he had lost his mind.

"Mr. Mellema." Maurits spoke again in Dutch, still ignoring me. "Even if you married this nyai, this concubine, in a legal marriage, she is still not Christian. She's an unbeliever! And even if she were Christian, you, sir, are still more rotten than Amelia Mellema-Hammers, more rotten than all the rottenness you accused my mother of. You, sir, have committed a blood sin, a crime against blood! Mixing Christian European blood with colored, Native, unbeliever's blood! A sin never to be forgiven!"

"Go!" I roared. He still ignored me. "Disturbing people's homes. You say you're an engineer, but you have no manners at all."

He still ignored me. I moved forward a step and he moved back half a step, as if to show his disgust at being approached by a Native. Then he spoke to his father again.

"Mr. Mellema, you now know what you really are."

He turned his back on us, descended the steps, got into the carriage. He did not glance back or say another word.

Your papa still stood nailed to the floor, in a state of stupefaction.

"So this is the child you had with your legal wife," I shouted at Tuan. "So this is the European civilization about which you've been teaching me all these years? You've been glorifying to the heavens? Night and day? Checking into people's private lives and homes, insulting them, so as to come and blackmail them? Blackmail? What else if not to blackmail? Why else does one investigate the affairs of other people?"

Ann, Tuan didn't hear my shouting. His eyeballs moved, but he gazed unblinking at the main road. I shouted at him again. He still didn't hear. Several workers came running up wanting to know what had happened. When they saw me berating Tuan so furiously, they all ran away.

I pulled at him and I scratched his chest. He was silent, and he didn't feel anything. But the hurt in my heart went amok, striking out looking for a target. I didn't know what he was thinking. Perhaps he was remembering his wife. How my heart hurt, Ann, especially because he didn't want to know about the pain I was suffering in this bosom of mine.

Tired of pulling and scratching, I began to cry. I collapsed, sitting down exhausted, like old clothes dropped on a chair. I lay my wet face down on the table.

When would my humiliation end? Must everyone be allowed to hurt me? Must I curse my dead parents, who had sold me as a nyai? I had never cursed them, Ann. Couldn't he understand, an educated person, an engineer, that he had not only humiliated me but also my children? Must my children become a rubbish bin, a place for insults to be flung? And why didn't my master, Mr. Herman Mellema, tall and big, broad-chested, hairy, and with mighty muscles, have my strength to defend his life-friend, the mother of his own children? What's the use of a man like that? He wasn't just my teacher, but the father of my children, my god. What was the use of his knowledge and learning? What was the use of his being a European respected by all Natives? What was the use of his being my master and my teacher at once, and my god, if he wasn't even able to defend himself?

From that moment on, Ann, my respect for him vanished. His teachings about self-respect and honor had become a kingdom within me. He was no better than Sastrotomo and his wife. If that was all he was able to do when faced with such a little test as this, then even without him, I could still look after my children, could still do everything myself. How my heart hurt, Ann; it's impossible that I could ever feel a greater pain as long as I live.

When I lifted my head, I saw him still standing there without blinking, stupefied, looking out towards the main road. Did he look at me, his life-friend and foremost helper? No, he didn't. He coughed, took a step, very slowly. He called out slowly, as if afraid of being heard by the devils and demons:

"Maurits! Maurits!"

He walked down the stairs, crossed the front yard. When he reached the main road, he turned right, towards Surabaya. He was not wearing shoes; he was in field clothes, wearing sandals.

Your papa didn't come home that day. I didn't care. I was still preoccupied attending to my pain. He didn't return home that night either. The next morning—still no sign. Three days and nights, Ann. During all that time the tears drenching my pillow were to no avail.

Darsam looked after everything. On the third day, he summoned up the courage to knock on the door. It was you, Ann, who opened the door and brought him upstairs. I never dreamed he'd dare come up. My hurt and sadness at once exploded into anger: How dare he come upstairs! Then it came to me: Perhaps he thought there was something more important than my pain and sadness. The door wasn't locked. It was you, Ann, who opened it. Perhaps you've forgotten it all now. That was the first and last time he came upstairs.

Darsam spoke like this:

"Nyai, except for the reading and writing, Darsam has looked after everything." He spoke in Madurese. I didn't answer. I wasn't thinking about the business. I was still lying on the bed, hugging my pillow. "Nyai shouldn't worry. Everything is fixed. You can trust Darsam, Nyai."

And it turned out he could indeed be trusted.

On the fourth day I went out from the house and garden. I took you from school. This business, the fruit of our efforts, must not collapse, must not be wasted. It was everything; upon it, our lives rode. It was my first child, Ann, your eldest brother, this business.

(Having finished telling her story, Mama wept and sobbed, feeling over again the pain of her humiliation, a humiliation she could neither answer nor revenge. After the crying died away, she resumed.)

You know yourself how many people—fifteen—I dismissed. It was they who had sold information to Maurits for small change. Perhaps they weren't even paid anything. And I must ask your forgiveness too, Ann. Your papa and I had agreed to send you to school in Europe, where you could study to be a teacher. I felt I sinned greatly when I took you from school. I've forced you to work so hard, long before you were old enough, to work every day, without holidays, without friend or comrade, because you mustn't have any, for the business's sake. I have made you learn to be a good employer. And employers may not become friends with their employees. You mustn't be influenced by them. What can one do, Ann?

After the arrival of Engineer Mellema, tremendous changes had occurred. I knew about Papa without anyone telling me. On the seventh day he returned. The strange thing was that he wore clean clothes and new shoes. That was in the evening, after the workday was over. Mama, Robert and I, were sitting at the front of the house. And Papa came.

"Don't speak to him. Don't greet him," Mama ordered.

The closer Papa came the more visible was his pale, cleanly shaven face. His hair was now parted in the middle. The odor of hair oil, never used at home, excited our nostrils. Also the smell of strong drink mixed with spices. He passed us without a remark or a glance, went up the steps, disappeared inside.

Suddenly Robert stood up and, with eyes popping out, stared at Mama and frowned angrily:

"My papa is not a Native!" He ran calling after Papa.

I looked at Mama. And Mama was looking at me. She said slowly:

"If you like, you may follow your brother's example."

"No, Mama," I exclaimed and I hugged her neck. "I only follow Mama, I'm a Native too, like Mama."

Mas, that's our true situation. I don't know if you're going to despise us like my brother, Robert, and Engineer Maurits, my stepbrother.

Who knows what Papa did in the house. We didn't know. All the rooms upstairs and downstairs were locked.

After a quarter of an hour he came out again. This time he looked at Mama and me. He offered no greeting. Behind him followed Robert. Papa left the yard again, descended to the main road, and disappeared. Robert came back to the house with a gloomy face, disappointed because Papa had ignored him.

I've hardly ever seen Papa since that event five years ago. Sometimes he appears, not saying anything, and then leaves, not saying anything. Mama refused to look for him or look after him.

Mama forbade me too to look for him. We were even forbidden to talk about him. Papa's portraits were taken down from the walls by Darsam and Mama ordered that they be burned in the yard in front of the whole household and all the workers. Perhaps that was how Mama let go of her feelings of revenge.

At first Robert was just silent. Only after Papa's portraits were burned did he protest. He ran inside the house, took down Mama's portraits, and burned them himself in the kitchen.

"He can join his father," Mama said to Darsam.

And the fighter passed on Mama's words to Robert, adding:

"Whoever dares disturb Nyai and Noni, I don't care if it's you yourself, Sinyo Robert, he will die under my machete. Sinyo can try if he likes, now, tomorrow, or whenever. And if Sinyo tries to find Tuan . . ."

Two months after all this, Robert graduated from E.L.S. He never reported it to Mama, and Mama didn't care either. He wandered about everywhere. A silent enmity has continued between Mama and Robert till this day. Five years.

At first Robert sold anything he could lay his hands on, from the warehouse, kitchen, house, office. He kept the money for himself. Mama got rid of every worker who took orders from him to steal. Then Mama forbade Robert from entering anywhere other than his own room and the dining room.

Five years passed, Mas. Five years. And there appeared two guests: Robert Suurhof for my brother, and Minke for me and Mama. Yes, you, Mas, no one else but you.

6

After I had been living for five days in that luxurious house at Wonokromo, Robert Mellema invited me to his room.

I entered warily. There was more furniture there than in my room. There was a desk with a glass top. Underneath the glass top was a big picture of a freighter, the *Caribou,* a ship flying the English flag.

He seemed friendly. His eyes were wild, a bit red. His clothes were clean and smelled of cheap perfume. His hair shone with pomade and was parted on the left. He was a handsome youth, tall, agile, fit, strong, and polite, and he looked as if he was always thinking. It was only his brown, marblelike eyes with their stealthy glances and his curled-up lips that really put me on edge. With just the two of us there I felt very uneasy.

"Minke," he began, "it looks as if you like living here. You're a school friend of Robert Suurhof, aren't you? In the same class at the H.B.S.?" I nodded suspiciously.

We sat on chairs, facing each other.

"I should have gone to H.B.S. too, and would have already graduated by now."

"Why didn't you go on?"

"That was Mama's responsibility, and Mama didn't do it."

"Pity. Perhaps you never asked her."

"No need to ask. It was her responsibility."

"Maybe Mama thought you didn't want to go on."

"There's no use in supposing about fate, Minke. This is my situation now. I'm outdone by you, Minke, you, just a Native— an H.B.S. student. But what's the point of talking about school?" He was silent a moment, examining me with his chocolate eyes. "I want to ask how you come to be staying here? It seems you like it here too. Because of Annelies?"

"Yes, Rob, because your sister is here. Also because I was asked."

He cleared his throat as I scanned his face.

"You've got objections perhaps?" I asked.

"You like my sister?" he asked in return.

"Yes."

"What a pity you're only a Native."

"It's a crime to be a Native?"

He cleared his throat once again, looking for words. His eyes wandered outside the window. I took the chance to check out his room more closely.

His bed had no mosquito net. In a hole in the floor there stood a bottle with the remnants of a mosquito coil in its neck. Around the bottle were scattered ashes. The room hadn't been swept yet.

I stopped inspecting the hole when I heard his voice again.

"This house is too quiet for me," he said. "Do you like playing chess?"

"Unfortunately no, Rob."

"Yes, a pity. Do you hunt? Let's go hunting."

"I am sorry, Rob, but I need the time to study. Actually, I like hunting too. Perhaps some other time?"

"Good, another time." He looked piercingly into my eyes. I knew there was a threat in that look. He dropped the palm of his right hand to his thigh. "How about if we go for a walk?"

"It's a pity, Rob, but I have to study."

We were both silent for quite a while. He stood up and closed the door. My eyes groped around seeking something to talk about while remaining ever vigilant, ready for every possibility. The window caught my attention. If he suddenly attacked me, that's where I'd run and jump out. Especially as below it there was a bench without any flowerpots.

On the other chair, the one Robert wasn't sitting on, lay a squashed, folded-up magazine. It looked as if it had been used as a prop for a cupboard or table leg.

"You don't have anything to read?" I asked. He sat in his chair again and answered with a voiceless laugh. His teeth were white, well looked after and gleaming.

"You mean by something to read, that paper?" His eyes pointed to the folded-up magazine. "I've had a look through it."

He picked it up and gave it to me. At that moment I was unsure if he wasn't about to do something. His eyes pierced my heart and I shivered. It was a magazine. Its cover was damaged, yet I could still read part of its name: *Indi*. . . .

"A magazine for lazy people," he said sharply. "Read it if you like. Take it."

From the paper and ink, you could tell it was a recent edition.

"What do you want to be when you've graduated from H.B.S.?" he suddenly asked. "Robert Suurhof says you'll be a bupati."

"It's not true. I don't want to be an official. I prefer being free, as I am now. And anyway, who'd make me a bupati? And you yourself Rob?" I asked in return.

"I don't like this house. I don't like this country either. Too hot. I like the snow better. This country is too hot. I'll go home to Europe. Sail. Travel the world. As soon as I board my first ship, I'll tattoo my chest and arms."

"That'd be great," I said. "I'd like to see other countries too."

"The same. So we could sail around the world together. You and me, Minke. We could make a plan, yes? It's a pity you're a Native. Look at the picture of this ship. A friend gave it to me." His spirits picked up. "He was a sailor on the *Caribou*. I met him by coincidence at Tanjung Perak. Spoke a lot, mainly about Canada. I was ready to join him. He wouldn't let me. 'What's the point in you becoming a sailor. You're the son of a wealthy man. Stay at home. You could buy your own ship if you wanted to.' " He gazed at me with dreaming eyes. "That was two years ago.

106

And he's never put in at Perak again. Hasn't sent any letters either. Perhaps he's drowned."

"Maybe Mama won't allow you to go," I said. "Who'll look after this big business."

"I'm an adult," he hissed, "with the right to decide for myself. But I'm still unsure. I don't know why."

"It's better you talk it over with Mama first." He shook his head. "Or with Papa," I suggested.

"A pity." He sighed deeply.

"I've never seen you talk with Mama. Perhaps it'd be all right if I told her?"

"No. Thank you. I hear from Suurhof you're a bit of a crocodile with the ladies."

I felt my face pound as my blood galloped through my veins. I knew at once I'd got to his sensitive spot. Now his true intentions were being revealed.

"Everyone has been judged good or bad by a third party at some time or other. Equally, on the other hand, everyone has passed judgment on other people too. I have. You have. Suurhof has," I said.

"Me? No," he answered firmly. "I've never cared about what other people say or do. Especially not about what people say of you. And even less so about what they say of me. But Suurhof said: Be careful of that filthy Native Minke, a 'low-class crocodile.' "

"He's right, everybody has a duty to be careful. Suurhof too. Nor am I any less careful of you, Rob."

"Look, I've never dreamed of sleeping in another person's house because of a woman. Even if I had been invited."

"I've already told you, I like your sister. Mama asked me to stay here."

"Good. As long as you know it wasn't me that invited you."

"I fully understand, Rob. I've still got Mama's letter."

"Let me read it."

"It's for me, Rob, not for you. A pity."

As time went on, his attitude and his voice increasingly showed his hatred. He was trying to frighten me.

"I don't know if you will marry my sister in the end or not. It seems Mama and Annelies like you. Even so, you must remember, I'm the male and the eldest child in this family."

He stammered and carefully scratched his head, afraid of messing up his hair.

"I know, you also know, all the people here are against me. Everyone ignores me. All this is *not* without its cause. Now you arrive. You're with them, no doubt. I stand alone here. It's best you never forget what a person standing alone can do," he said threateningly, with smiling lips.

"Yes, Rob, and don't forget your own words either, because they're directed at yourself as well." His eyes now dreamily gazed at me as he took the measure of my strength, and I followed his example, also smiling. I followed all his movements. At the slightest indication of a suspicious move, I'd be up and out the window. He would no longer find me in the room.

"Good," he said nodding. "And don't you forget either, you're only a Native."

"Oh, I'll certainly always remember that, Rob. Don't worry. Don't you forget either, in your veins runs Native blood too. I'm indeed not an Indo, not a Mixed-Blood European; but while I'm studying at European schools, there's European knowledge and learning inside me too, if it's European things that you value so much."

"You're clever, Minke, fit to be an H.B.S. student."

That short conversation felt as if it had gone on tensely for hours. Then I realized it had lasted only ten minutes. Luckily Annelies called from outside, and I excused myself.

Catching me entirely by surprise, Robert, still sitting, said calmly:

"Go, your nyai is looking for you."

I stopped at the door and looked at him in astonishment. He only smiled.

"She's your sister, Rob. You shouldn't talk like that. I too have my honor . . ."

Annelies hurriedly pulled me away to the back room as if something important had occurred there. We sat on the thick-cushioned sofa. The cover had a floral pattern on a cream base. She clung to me and whispered.

"Don't see too much of Robert, let alone go into his room. I worry. Each day he changes more. Twice now Mama has refused to pay his debts, Mas."

"Do you have to be enemies with your own brother?"

"It's not that. He must work to earn his living. He could if he wanted to. But he doesn't want to."

"Yes, but why must you two be enemies."

"It doesn't come from my side. Mama is right in everything. He doesn't want to acknowledge that Mama is right just because she is a Native. So what must I do?"

I didn't press her further. But I thought to myself then: What was that handsome youth getting from this family? From his mother nothing, from his father nothing. Let alone his sister. Neither sympathy nor love. I come to his home, and he's jealous of me. It's only natural.

"Why don't you try to be the peacemaker, Ann?"

"What for? He's gone too far, made me curse him."

"Curse him? You've cursed him?"

"I don't want to even look at his face. Before I was still prepared to be good to him. Now, for as long as I live—never. Never, Mas."

I regretted having tried to interfere. And her face, which all of a sudden reddened, showed her anger.

Nyai joined us. She had a copy of the *Surabaya Daily News* in her hand. She pointed out to me a short story, "Een Buitenge-woon Gewoone Nyai die Ik ken," ("An Extraordinary Ordinary Nyai That I knew").

"Have you read this story, Nyo?"

"Yes, Mama, at school."

"I think I recognize the person described in this story."

Perhaps I went pale on hearing her words. Although the title had been changed, that was my own writing, my first short story published in other than an auction paper. A few words and sentences had been improved, but it was still my work. The material wasn't from Annelies, but my own imaginings based on Mama's day-to-day life.

"Who's the author, Mama?" I asked, pretending.

"Max Tollenaar. Is it true you only write advertisements?"

Before the conversation became involved I quickly confessed: "Yes, I wrote it, Ma."

"I thought so. You're indeed clever, Nyo. Not one in a hundred people can write like that. Though if you mean me in that story . . ."

"I based my imaginings on Mama," I answered quickly.

"I see. It's not surprising then that there are a lot of incorrect things in it. As a story it's indeed good, Nyo. Let's hope you become a writer, like Victor Hugo."

And I was embarrassed to ask who Victor Hugo was. And she could praise the strengths of a story. When did she ever study story-writing? Or is it all just pretense?

"Have you read Francis? G. Francis?"

I felt truly at a loss. I'd never heard of him either.

"It seems Sinyo never reads Malay."

"Malay books, Mama? Are there such things?"

"It's a pity you don't know. He's written many Malay books. I think he's a Pure or a Mixed-Blood, not a Native. It would be a real pity, Nyo, if you never read any of his books."

She spoke still more about the world of fiction. And the more she spoke, the more I doubted her. She might be just parroting everything ever taught to her by Herman Mellema. My teachers taught us quite a lot about Dutch language and literature. They had never touched on the things that she was discussing. And my favorite teacher, Miss Magda Peters, must know more than any nyai. This nyai was even trying to talk about literary language as well!

"Francis wrote *Nyai Dasima,* a truly European-style novel. But in Malay. I've got the book. Perhaps you'd like to study it."

I mechanically replied yes. What did she know about the literary world? And why did she want to read stories, interfere in the affairs of the imaginary characters of the world of writers, even commenting on the language they used, while under her own eyes her own son Robert was neglected? It gave me cause to have doubts.

As if she could read my thoughts, she said:

"Perhaps you want to write about Robert too."

"Why, Mama?"

"Because of your youth. You'll want to write, of course, about the people you know around you. Who interest you. Who excite your sympathy or antipathy. I think Rob will certainly attract your attention."

Luckily that rather unpleasant conversation was soon followed by dinner. Robert didn't join us. Both Mama and Annelies were not surprised, nor did they inquire after him. The servant didn't ask any questions either.

In the middle of dinner I was about to tell them of Robert's desire to become a sailor, to go home to Europe. At that moment Nyai spoke instead.

"Write, Nyo, always about humanity, humanity's life, not humanity's death. Yes, whether it's animals, ogres, gods, or ghosts that you present, there's nothing more difficult to understand than humanity. That's why there is no end to the telling of stories on this earth. Every day there are more. I don't know a lot about this myself. But once I read something that said, more or less, the following:

Do not underestimate the human being, who sometimes appears so simple. Even with sight as sharp as an eagle, a mind as sharp as a razor, senses more powerful than the gods, hearing that can catch the music and the lamentations of life, your knowledge of humanity will never be total.

Mama had stopped eating altogether. The spoon with its load still hovered in midair under her chin. "It's true that during the last ten years I've read more fiction. It's as if every book is concerned with people's efforts and striving to escape or overcome some difficulty. Stories about happy things are never interesting. They are not stories about people and their lives, but about heaven, and clearly do not take place on this earth of ours."

Mama resumed eating. I had concentrated all my attention in order to catch every one of her words. At that time she was really an unofficial teacher whose lessons were delivered officially enough.

After dinner she resumed:

"Because of all this, you are no doubt interested in Robert. He always seeks trouble and then can't get out of it. I think, if I'm not wrong, that's what's called tragic. The same as his father. Perhaps through your writings—if he was willing to read them— he could see himself as in a mirror. Perhaps he could change his ways. Who knows? Only, I ask, before you publish them, let me see them first. Of course, that's if you don't mind. You never know but that there might be some false impressions or assumptions that could be avoided."

*   *   *

111

As it happened I was indeed working on a piece about Robert. Nyai's warning rather startled me. It was if her eagle eyes saw my every move. I also felt that she was invading my privacy as a storyteller. The publication of my first story had raised my spirits. But the enthusiasm that success had generated could not now spur on my writings about Robert. Mama, with her eagle eyes, had blocked that progress in the middle of its journey.

The discussion at dinner made me submerge myself in reflections. It was clear that she had read a great deal. Probably Mr. Herman Mellema was a truly wise and patient teacher. Nyai was a good pupil, and she had the ability to develop herself, after having obtained some capital in the form of understanding from her master. What I can't obtain from school, I will harvest in the midst of this concubine's family. Who would ever have guessed? Perhaps, too, she does understand Robert Mellema better than I can. What she had said about the Native-hating youth pointed to a deep compassion for her eldest child.

I still didn't know much about that tall youth. Perhaps he too read a lot, like his mother. The magazine he gave me turned out to be no ordinary magazine but a copy of a well-known scholarly historical journal. It could have come from Mama's library, or have been taken from the postman and not handed over to Mama. Maybe, too, he hadn't finished reading it either. I don't know for sure. All the articles in it were about the Netherlands Indies. One among them was about Japan's relations—whether they are limited or extensive—with the Indies.

That article greatly enriched my notes about Japan—a country that had been very much discussed during those preceding few months. None of my school friends had paid any attention to that country even though it had been touched upon in discussions at school. My friends still did not consider Japan as worthy of discussion. They offhandedly equated Japan with the prostitutes who filled up the Kembang Jepun, and with the little cafés, restaurants, and barber shops, with the hawker and his goods. None of these reflected the Japan that was challenging modern science and learning.

In one discussion, when my teacher, Mr. Lastendienst, tried to get the students interested, most just chatted lazily to each other. He said that Japan was also experiencing a flowering in the field of science. Kitazato had discovered the plague bacteria, Shiga had discovered dysentery bacteria—and in that manner Japan, too, had been of service to humanity. He compared it with the Dutch nation's contribution to civilization. Seeing that I was fully engaged in the subject and was taking notes, Mr. Lastendienst asked me in an accusing tone of voice, "Eh, Minke, the Javanese delegate in this room, what has your nation contributed to humanity?" I would not have been alone in being so startled to hear that sudden question; in all likelihood all the gods in the chest of the shadow play puppet-master would have exhausted their energy just to answer. So the best way of getting out of my difficulty was to utter the following sentence: "Yes, Mr. Lastendienst, I can't answer at this time." And my teacher reacted to this with a sweet smile—very sweet.

That's just a little from my notes about Japan. Now with the articles in the magazine Robert gave me, my notes had been supplemented by quite a bit of extra information about the current developments in Japan and the struggle over its defense strategy. I didn't understand much about those things. Precisely because of that I noted it all down. At the very least it would be excellent material for use in a school discussion.

It said there had been competition between the Japanese Army and Navy. A maritime defense strategy was then chosen. And the army, with its centuries-old samurai tradition, was dissatisfied.

And the Indies itself? In the article it said the Netherlands Indies has no navy, only an army. Japan is made up of islands. The Netherlands Indies is just a great string of them. Why does Japan emphasize naval defense while the Indies emphasize the land? Isn't the problem of defense (against the outside) the same? Didn't the Indies fall into the hands of the English a hundred years ago precisely because of the weakness of the Indies Navy? Why hasn't that lesson been learned?

The warships that sailed back and forth in Indies waters did not belong to the Netherlands Indies but to the Kingdom of the Netherlands. Governor-General Daendels had made Surabaya a navel base in a period when he had not a single ship! Almost a hundred years later still no one gave any thought to the Indies having its own navy. The honorable gentlemen in charge put their trust in the British naval defenses of Singapore and the American naval defenses of the Philippines.

The article speculated about war with Japan. In what kind of

113

position would the Indies be with its undefended waterways? While the Royal Dutch Navy carried out only irregular patrols? Could not the experience of 1811, when the Indies fell into British hands, be repeated once again to Holland's loss?

I don't know whether Robert ever read and studied the article. Seeing that he wanted to travel the world as a sailor, perhaps he had read it. And as someone who believed in European superiority, he would no doubt put his faith in the white race. The article also said that Japan was trying to imitate the English on the seas. And the writer warned people that they should stop insulting the Japanese by calling them imitating monkeys. At the beginning of all growth, everything imitates. All of us, when we were children, also only imitated. But children grow up and begin their own development.

And the same theme occurred in a conversation between Jean Marais and Telinga that I once overheard, and I noted it down like this:

Jean Marais: Roles shift from one generation to another, from one nation to another. Previously, colored peoples conquered the white peoples. Now the white peoples conquered the colored peoples.

Telinga: The whites had never been beaten in three centuries by coloreds. Three centuries! Indeed it could happen that one white people could conquer another white people. But a colored people could never conquer whites. Not in the next five centuries. Not ever.

And Robert wanted to become a sailor, a European. He dreamed of sailing on the *Caribou,* under the flag of England—a country of no great size, with a sun that never set.

7

It felt as if I hadn't been asleep for long. Nervous knocking on the door jolted me awake.

"Minke, wake up." It was Nyai's voice.

I found Mama standing at the door, carrying a candle. Her hair was a bit of a mess. In the morning darkness the tick-tock of the pendulum clock reigned over the room.

"What's the time, Mama?"

"Four. Someone is looking for you."

A person was sitting on the settee in the gloom. The closer Mama's candle came, the clearer the person became: a police officer! He stood up out of respect, then immediately spoke in Malay but with a Javanese accent:

"Tuan Minke?"

"Yes."

"I have an order to take Tuan with me. Straight away." He held out a letter. What he said was true. It was a summons from the police station at B——, a town near the town of my birth, approved by the police station at Surabaya. My name was clearly stated there. Mama too had read it.

"What have you been up to, Nyo?" she asked.

"Not a thing," I answered nervously. Yet I began to have doubts about my own actions. I searched and searched my memory; I lined up everything I'd done over the last week. I repeated, "Not a thing, Mama."

Annelies came. She wore a long, black-velvet gown. Her hair was a mess. She was still bleary-eyed.

Mama approached me:

"The officer hasn't said what you've been charged with. It's not in the summons either." And to the police officer: "He has the right to know what is the matter."

"I have no such orders, Nyai. If it is not stated clearly in the summons, then indeed no one may be told, including the person involved."

"It can't be done like that," I retorted. "I'm a Raden Mas, I can't be treated in this way," and I waited for an answer. Seeing that he didn't know how he should answer, I resumed, "I have *forum privilegiatum,* the right to be tried under the same laws and in the courts of the Dutch."

"No one can deny that, Tuan Raden Mas Minke."

"Why are you doing things in this manner?"

"My orders are only to fetch Tuan. Even the person who issued the orders would not know anything more, Tuan Raden Mas," he said, defending himself. "Please get ready, Tuan. We must leave quickly. We must be at our destination by five o'clock this afternoon."

"Mas, why do they want to take you?" asked Annelies, afraid. I sensed a shiver in her voice.

"He won't say," I answered briefly.

"Ann, fix Minke's clothes and bring them here," ordered Nyai, "Who knows how long they will detain him. He can bathe and breakfast first, can't he?"

"Of course, Nyai, there's still a little time."

He allowed half an hour.

I found Robert watching the event from his room at the back. A yawn was his only greeting. Once in the bathroom, I began to mull over the possibilities: Robert is the cause of this trouble, passing on false and fanciful reports. He didn't appear for dinner last night or the night before. One by one his threats came back to me. All right, if it's true you are the one who has caused all this disturbance, I will never forget you, Rob.

On my return to the front room I found coffee and cakes ready. The police officer was enjoying his breakfast. He looked more polite after receiving his food. And he didn't appear to have any personal enmity towards us. On the contrary, he chatted to us, laughing all the time:

"Nothing bad is going to happen, Nyai," he said finally. "Tuan Raden Mas Minke will return, at the very latest, within the next two weeks."

"It's not a matter of two weeks or a month. He's been arrested in my house. I have a right to know what it's all about," pressed Nyai.

"Really, I don't know. Forgive me. That's why I'm fetching him so early in the morning, Nyai, so no one will know."

"So no one will know? How? You had to meet with my watchman before getting to see me, didn't you?"

"Then arrange so that your watchman doesn't talk."

"You can't do this to me," said Mama. "I'm going to ask the police station for an explanation."

"That's a good idea. Nyai will get an explanation quickly. And it will surely be a truthful one."

Annelies, who was standing holding my suitcase, approached me, unable to speak. She put the suitcase and bag down. She grabbed hold of my hand and held it. Her hand trembled.

"Breakfast first, Tuan Raden Mas," the officer reminded me. "Maybe there won't be any breakfast as good as this at the police station. No? Then let's leave now."

"I'll be back soon, Ann, Mama. There must be some mistake. Believe me."

And Annelies wouldn't let go of my hand.

The police officer picked up my things and carried them outside. Annelies gripped my hand tightly as I followed him out. I kissed her on the cheek and freed myself of her hold. And she still didn't speak.

"Hopefully all will be well, Nyo," Nyai prayed. "That's enough now, Ann; pray for his safety."

The carriage that awaited us proved not to be a police carriage but a hired one. We climbed aboard and left in the direction of

117

Surabaya. The officer was taking me to B——. And in the early morning dark I conjured up in my mind all the buildings I'd ever seen in B——. Which one among them was our destination? The police station? The jail? An inn? It didn't even occur to me to think of private houses.

Our carriage made up the only traffic on the road. Oil wagons, which usually started moving out of the D.P.M. refinery at dawn, twenty or thirty in a row, were not yet to be seen. One or two people were carting vegetables on their backs for sale in Surabaya. And the agent kept his mouth closed as if he'd never, in all his life, learned to speak.

Maybe it was true that Robert had slandered me. But why was B—— our destination?

Our oil lamps were reluctant to pierce the darkness of the misty early morning. It was as if only I, the policeman, the driver, and the horse were alive upon that road. And I imagined Annelies crying unconsoled. And Nyai confused, concerned that my arrest would give her business a bad name. And Robert Mellema would have reason to cackle: See! Isn't it true what Suurhof said?

The carriage took us to the Surabaya police station. I was invited to sit down and wait in the reception room. I wanted very much to ask questions about my case. But I got the impression that in the misty, early-morning air no one was in the mood to give explanations. So I didn't ask. And the carriage still waited in front of the police station. The officer just left me there alone with no orders.

It was a long time. The sun still hadn't risen. And when it did rise, it was unable to dispel the mist. Those gray particles of water reigned over everything, even the inside of my lungs. The traffic in front of the police station was beginning to get busy: carriages, carts, pedestrians, hawkers, workers. And I still sat alone in the waiting room. Finally, at a quarter to nine, the police officer appeared. He looked as if he'd had an hour's sleep and had had a hot bath. He looked fresh. On the other hand, I felt weary, exhausted by waiting. And still I had no chance to inquire.

"Let's go, Tuan Raden Mas," he invited, in a friendly manner. We climbed back aboard the carriage and headed off to the railway station. Once again it was he who carried my bags, and unloaded them and escorted me to the ticket window. He pushed a letter inside and obtained two tickets—first class. This was not the time for the express train. We were traveling on a slow train too. Yes, it was true, we climbed aboard that boring train on the western line.I'd never been on a train such as this. I always went by express if there was one. Except, yes, except from B—— to my own town T——.

The officer returned to his state of muteness. I sat next to the window. He sat facing me.

There were just a few passengers in the carriage. Besides the two of us, there were three European men and a Chinese man. They all seemed bored. At the first stop two passengers got off, including the Chinese man. No new passengers came aboard.

I'd traveled that distance dozens of times. So there was nothing in the sights along the journey that interested me. In B—— I usually stayed at an inn until the next morning, when I would continue the journey on to T——. This time I wouldn't be heading towards my usual inn. Most probably I would end up at the police station.

The view became more and more dreary: barren land, sometimes gray, sometimes a whitish-yellow. I fell asleep with a hungry stomach. Whatever was going to happen, let it. Ah! This earth of mankind. Sometimes a tobacco plantation would appear, shrink, and disappear, swept away in the train's acceleration. Appear again, shrink again, disappear again. And paddy fields and paddy fields and paddy fields, unirrigated, planted with crops, but no rice, almost ready to be harvested. And the train crawled on slowly, spouting thick black and dusty and sparking smoke. Why wasn't it England that controlled all this? Why Holland? And Japan? What about Japan?

The touch of the police officer's hand woke me. Laid out beside me were the things he'd brought with him: the wrapping cloth was opened out as a kind of tablecloth. On it was fried rice shining with oil, adorned by a fried egg and fried chicken, plus a spoon and fork, all in a banana-leaf container. Perhaps it had been specially prepared for me. An agent would think twice about serving food like this for himself—it was too sumptuous. A white bottle containing chocolate milk—a drink not yet widely known by Natives—stood beside the banana-leaf container.

And that gloomy town, B——, finally, as the time approached five o'clock in the afternoon, appeared before our eyes. The agent still didn't speak. But again he carried my bags. And I didn't stop him. What was the significance of a police officer, first-class, compared with an H.B.S. student? At the most he might have been able to read and write a little Javanese and Malay.

A carriage took us away from the station. To where? I knew those white, rocky streets that were such an eyesore. Not to the hotel, not to my regular inn. Also not to the B—— police station.

The town square looked deserted with its brownish grass carpet, balding and tattered in places. Where was I being taken? The hired carriage headed for the bupati's residence and stopped a distance from its stone gate. What was the link between my case and the bupati of B——? My mind began groping around madly.

And the police officer alighted first, looking after my bags as before.

"Please," he said suddenly in high Javanese.

I accompanied him into the regency office situated diagonally across from the bupati's residence. An office stripped of wall ornaments, devoid of appropriate furniture, without a single occupant. All the furniture was rough, made from teak, and unvarnished, with the appearance of not having been measured for need and without any plan of use, just thrown together. Coming from the luxurious house in Wonokromo into this room was like making a visit to a produce house. It was, you might say, just slightly more luxurious than Annelies's chicken coop. This, presumably, was the interrogation room. Just some tables, a few chairs, and some long benches. On the other side, there were shelves on which were piles of papers and several books. There were no instruments of torture. Just ink bottles on all the tables.

The police officer left me by myself again. And for the second time I waited and waited. The sun had set. He still didn't appear. The grand mosque's drum began its beating, followed by that sad call to prayer. The street lanterns were being lit by the lamp men. The office became darker. And those demonic mosquitoes, they ganged up and attacked the room's only occupant. The insolence! I swore. Is this how people treat a raden mas and an H.B.S. student too? An educated person with the blood of the kings of Java running in him?

And I could feel my clothes sticking to my body. And my body was starting to stink of sweat. I'd never experienced such maltreatment as this.

"A thousand pardons, Ndoro Raden Mas." The officer invited me to leave the mosquito-filled, dark office. "Allow your servant to escort you to the visitors' gallery."

Once again he carried my bags.

So I'm being brought before the bupati of B——! God! What's it all about? And must I, an H.B.S. student, cringe in front of him and at the end of every one of my sentences, make obeisance to someone I don't even know? As I walked along the path to the visitors' gallery, already lit up by four lamps, I felt like crying. What's the point in studying European science and learning, mixing with Europeans, if in the end one has to cringe anyway, slide along like a snail, and worship some little king who is probably illiterate to boot. God, God! To have audience with a bupati is to be an object of humiliation without being able to defend oneself. I'd never forced anyone to act like that towards me. Why did I have to do so for others? Thundering damnation!

Ah, see, it's true? The agent was already inviting me—the impudence!—to take off my shoes and socks. The beginning of a great tyranny. Some supernatural power forced me to follow his orders. The floor felt cold under my bare soles. He signaled me and I went, step by step, to the top. He pointed out to me the place where I must sit, eyes towards the floor, before a rocking chair. One of my teachers had once said: The rocking chair is the most beautiful thing left behind by the Dutch East Indies Company after its bankruptcy. Oho! Oh rocking chair, you will be a witness to how I must humiliate myself in order to glorify some bupati I don't even know. Damn! What would my friends say if they saw me traveling on my knees like this, like someone without thighs, crawling towards the relic of the company at the time it approached bankruptcy—that unmoving chair near the wall of the visitors' gallery.

"Yes, walk on your knees, Ndoro Raden Mas." The officer was herding a buffalo into a mudhole.

And I covered the almost ten meters distance while swearing in three languages.

To my left and right clam-shell ornaments were spread out. And the floor shone from the rays of light from four oil lamps. Truly, my friends would ridicule me if they could see this play, where a human being, who normally walks on his two whole legs, on his own feet, now has to walk with only half his legs, aided by his two hands. *Ya Allah!* You, my ancestors, you: What is the reason you created customs that would so humiliate your own descendants? You never once gave it any thought, you, my ancestors who indulged in these excesses! Your descendants could have been honored without such humiliation! How could you bring yourself to leave such customs as a legacy?

I stopped in front of the rocking chair. I sat, legs tucked under and eyes towards the floor, as custom decreed. All I could see was a low, carved bench and on top of it, a foot-rest cushion of black velvet. The same velvet as Annelies's dressing gown earlier that morning.

Good, now I had sat down cross-legged before that damned rocking chair. What business did I have with the bupati of B——? None. Neither kith nor kin, not an acquaintance, let alone a friend. And for how much longer would this oppression and humiliation continue? Waiting and waiting while being oppressed and humiliated in this way?

I heard the creaking of a swinging door as it opened. Then, as the seconds passed, the sound of footsteps made by leather slippers became clearer. And I remembered the scraping footsteps of Mr. Mellema on that other frightening night. From where I was seated, the striding slippers slowly began to come into view. Above them a pair of clean legs. Still further above a widely pleated batik sarong.

I raised my hands, clasped in obeisance, as I had seen the court employees do before my grandfather, and my grandmother, and my parents at the end of Ramadan. And I did not now withdraw my pose until the bupati had sat himself comfortably in his place. In making such obeisance it felt as if all the learning and science I had studied year after year was lost. Lost was the beauty of the world as promised by science's progress. Lost was the enthusiasm of my teachers in greeting the bright future of humanity. And who knows how many times I'd have to make such obeisances that night. Obeisance—the lauding of ancestors and persons of authority by humbling and abasing oneself! Level with the ground if possible! I will not allow my descendants to go through such degradation!

This person, the bupati of B——, cleared his throat. Then slowly he sat down on the rocking chair, kicking off his slippers behind the foot bench, and placed his honorable feet on the velvet cushion. The chair began to rock a little. Damn! How slowly time passed. Some object, by my reckoning fairly long, gently tapped upon my uncovered head. How insolent was this being that I must honor. And every tap I must greet with a sign of grateful obeisance.

After five taps, the object was withdrawn, and was now hung beside the bupati's chair: a horse whip made from a bull's genitals, with a shaft of thin, choice leather.

"You!" he addressed me weakly, hoarsely.

"Yes, I, my master, Honored Lord Bupati," said my mouth, and like a machine my hands were raised in obeisance for the umpteenth time and my heart cursed for I don't know how many times now.

"You! Why have you only come now?" His voice now emerged more clearly from his throat, which was suffering the end of a bout of influenza.

It felt as if I had heard that voice before. It was also his cold which prevented me from recalling the voice easily. No, no, it could not be him! Impossible! No! I still didn't know what this was all about, so I kept silent.

"The honored government does not run a postal service for nothing—capable of getting my letters safely to you at the proper and exact address."

Yes, it was his voice. Impossible! It couldn't be! Impossible. I was just guessing.

"Why are you silent? Now that your schooling is so advanced, is it that you feel humiliated to have to read my letters?"

Yes, it was his voice! I raised my hands in obeisance once again, deliberately lifting my head a little and taking a peep. *Ya Allah!* It was indeed he.

"Father!" I cried. "Forgive me, your servant."

"Answer! You feel humiliated to have to reply to my letters?"

"A thousand pardons, my father: no."

"Your mother's letter, and why didn't you answer that either?"

"My father, a thousand pardons . . ."

"And the letter from your elder brother?"

"Forgive me, my father, a thousand pardons, I was not there. I'm no longer at that address, forgive me, a thousand pardons."

"So you were educated as high as a coconut tree to learn to deceive?"

123

"A thousand pardons, my father."

"You think we're all blind, ignorant of the date you moved to Wonokromo? And do not know that you took with you all our letters still unread?"

The horse whip made from bull's genitals swayed to and fro. The hairs on my back crawled, ready to receive the whip as it fell upon me, as a rebellious horse.

"Do you still need to be humiliated in public with this whip?"

"Humiliate me with the horse whip in public," I answered recklessly, unable to stand such tyranny. "But it would be an honor if that order were to come from a father," I continued, still more recklessly. And I would show the same attitude as Mama did to Robert, Herman Mellema, Sastrotomo, and his wife.

"Crocodile!" he hissed angrily. "I took you out of the E.L.S. Dutch-language primary school at T——for the same reason. As young as that! The higher your schooling, the more you turn into a crocodile! Bored of playing around with girls of your own age, you're now holing up with a nyai's nest. What do you want to become of you?"

I kept silent. Only my heart shouted in anger: So you insult me thus, blood of kings! Husband of my mother! Good, I will not answer. Come on, keep going, continue, blood of the kings of Java! Yesterday you were just an irrigation official. Now all of a sudden you are a bupati, a little king. Strike me with your whip, king, you who know not that science and learning have opened a new era on this earth of mankind!

"Prepared by your grandfather to be a bupati, to be honored by all people, the cleverest child in the family . . . the cleverest in the town . . . yes, God, what will become of this child!"

Good, come on, continue, little king!

"The only grounds for forgiving you are because you've passed and gone up a class."

I can go on up to eleventh class! I roared inside my breast, offended. Come on, let fly with your ignorance, little king.

"Don't you think it's dangerous to take up with a nyai? If her master goes into a rage and you're shot dead by him, or perhaps attacked with a dagger, or a sword, or a kitchen knife, or strangled . . . how will it be? The papers will announce who you are, who your parents are. What sort of shame will you bring upon your parents? If you haven't thought things through as far as that . . .''

Like Mama, I was ready to leave all my family, I roared louder inside, a family that burdens me with nothing but bonds that enslave,! Come on, continue, blood of the kings of Java! Continue! I too can explode.

"Haven't you read in the papers that tomorrow night your father is celebrating his appointment as a bupati? Bupati of B——? Mr. Assistant Resident of B——, Mr. Resident of Surabaya, Mr. Controller, and all the neighboring bupatis will be present. It is possible an H.B.S. student doesn't read the newspapers? If not, is it possible nobody else told you about it? Your nyai, can't she read the papers for you?"

Indeed the civil service reports were something that never attracted my interest: appointments, dismissals, transfers, pensions. Nothing to do with me. The world of priyayi, Javanese aristocrats who became administrators for the Dutch colonial bureaucracy, was not my world. Who cared if the devil was appointed smallpox official or was sacked dishonorably because of embezzlement? My world was not rank and position, wages and embezzlement. My world was this earth of mankind and its problems.

"Listen, you renegade!" he ordered, a newly important official whose spirits were now aroused. "You've become absentminded, looking after someone else's nyai. You've forgotten your parents, your duties as a child. Perhaps you're indeed ready to take a wife. All right, another time we'll discuss it. Now there is another matter. Pay attention. Tomorrow night you will act as interpreter. Don't shame me and the family in public before the resident, assistant resident, controller, and neighboring bupdtis."

"Yes, my father."

"You're able and ready to be interpreter?"

"Able and ready, my father."

"Ah, that's better, once in a while please your parents' hearts. I'd begun to worry that Mr. Controller would be carrying out that task. Imagine how it would look at a party to celebrate my appointment, with all the important officials as witnesses, if there was a son missing? When should they start becoming acquainted with you? This will be the best opportunity. It's a pity you're such a renegade. Perhaps you don't understand that your parents are clearing the way to a high position for you. You, a son, glorified as the cleverest in the family. Or perhaps you're more inclined towards the nyai than towards rank?"

"Yes, my father."

"This is how your road to high rank will be clear."

"Yes, my father."

"There, go to your mother. You indeed did not intend to return home. It was so shameful, having to ask the help of my assistant resident. You're happy, aren't you, being arrested like an unpracticed thief? No sense of shame at all. Go kneel before your mother, though I know you're resolved to forget her. Sever your relations with that nyai who doesn't know when she's already well off!"

Naturally, I did not answer. I just made the sign of obeisance again. Then, walking on half legs, assisted by my hands, I crawled off, carrying the burden of my indignation on my back, like a snail. Destination: the place I had taken off my shoes and socks, the place where this accursed experience began. There were no Natives in the bupati's building wearing shoes. With my shoes in my hands I walked alongside the visitors' gallery, entering the inner courtyard. The gloomy lanterns showed the way to the kitchen. I collapsed into a broken-down lounge chair, ignoring the things I was carrying.

Someone came to have a look. I pretended not to notice. I was served a cup of black coffee, which I gulped down.

If my elder brother hadn't turned up, perhaps I would have fallen asleep. Putting on a vicious countenance, he spoke to me in Dutch.

"It seems you've forgotten politeness too, and so have not gone quickly to kneel before Mother?"

I rose and accompanied him, a S.I.B.A. student, a future Netherlands Indies civil servant. He kept frowning as if he were the guardian whose job it was to ensure the sky wouldn't fall in and smash up the earth. Because his Dutch was limited, he resumed in Javanese his lecture about how I was a child who no longer knew proper custom. I, of course, didn't respond. We entered the bupati's building, passing by several doors. Finally, in front of one door, he said:

"Enter there, you!"

I knocked slowly on the door. I didn't know whose room it

was, but opened it and entered. Mother was sitting in front of the mirror combing her hair. A tall oil lamp stood on a stand beside her.

"Mother, forgive me," I said, kneeling down before her and kissing her knees. I don't know why my heart was seized so suddenly by this longing for my mother.

"So you have come home at last, Gus. Thank God you're safe." She lifted up my chin, looked into my face, as if I were a four-year-old child. And her soft, loving voice moved me. My eyes overflowed with tears. This was my mother, just as before, my own mother.

"This is Mother's wayward child," I submitted hoarsely.

"You're a man now. Your mustache is beginning to come through. People say you like a rich and beautiful nyai." Before I could deny this, she continued, "It's up to you, if you indeed like her and she likes you. You're an adult now. You're no doubt ready to shoulder the consequences and responsibilities and not run like a criminal." She took a breath and stroked my cheek as if I were a baby. "Gus, they say you are doing very well at school. Thanks be to God. It amazes me sometimes how your schooling can go so well while you're in the power of that nyai. Or perhaps you're truly very clever? Yes, yes, that's a male for you; all men are cats pretending to be rabbits. As rabbits they eat all the leaves, as cats they eat all the meat. All right, Gus, you must do well at school, keep advancing."

Mother didn't fault me. I did not have to deny anything.

"Men, Gus, they love to eat. Who knows if leaves or if meat? That's all right, providing you understand, Gus, the more you advance at school does not mean the more you can eat other people's food. You must be able to recognize limits. That's not too hard to understand, is it? If people don't recognize such limits, God will make them realize in His own way."

Ah, Mother, how many pearllike words have you burned into my soul.

"You're still silent, Gus. What are you going to report to your mother? My waiting is not going to be in vain, is it?"

"Next year I will graduate, Mother."

"Thanks be to God, Gus. Parents can only pray. Why have you only come now? Your father was so worried, Gus, angry every day because of you. Your father was appointed bupati very, very suddenly. No one guessed it would be so fast. You, one day, will reach the same heights. You surely must be able to. Your father only knows Javanese, you know Dutch; you are an H.B.S. student. Your father only went to a Basic People's School. You have mixed widely with the Dutch. Your father hasn't. You will surely become a bupati one day."

"No, Mother, I don't want to become a bupati."

"No? Strange. Yes, as you wish. So what do you want to become? If you graduate you can become whatever you want, of course."

"I only want to become a free human being, not given orders, not giving orders, Mother."

"Ha! Will there be a time like that, Gus? This is the first I've heard of it."

When I was a little boy, I used to tell her excitedly about what my schoolteachers had said. This time too. About Miss Magda Peters, whose stories were so interesting, about the French Revolution, its meaning, its basic principles. Mother only laughed, not refuting anything. Just as when I was a little child.

"Ugh! You're so dirty, you smell of sweat. Bathe, and with hot water! It's already so late. Rest. Tomorrow you'll be working hard. You know your duties tomorrow?"

I was not yet acquainted with the building. I went into the room prepared for me. An oil lamp was alight inside. It appeared that my brother was also in that room. He was sitting reading by the table lamp. I passed by to get my things ready. And my brother, who always exercised his rights as the firstborn, did not lift his head at all, as if I did not exist on this earth. Was he trying to impress me with his diligence as a student?

I coughed. He still showed no reaction. I glanced at what he was reading. Not printing: handwriting! And I became suspicious when I looked at the book's cover. Only I owned a book with a beautiful cover, hand-made by Jean Marais. Slowly I moved up behind him. I was not wrong: my diary. I seized it from him and I became enraged.

"Don't touch this! Who gave you the right to open it? Is this what your school has taught you?"

He stood up, staring at me wide-eyed, and said, "Indeed you are no longer Javanese."

"What's the use of being Javanese only to have one's rights

violated? Perhaps you don't understand that notes like that are very personal? Haven't your teachers taught you about ethics and individual rights?"

My brother was silent, observing me in a powerless anger.

"Or is this indeed the practice given to trainee officials? Fiddling in other people's affairs and violating the rights of anybody they like? Aren't you taught the new civilization? Modern civilization? You want to become a king who can do as he pleases, like your ancestors' kings?"

My resentment and anger had spilled out.

"And is this what the new civilization means? To insult people? To insult government officials? You yourself will become one!" he defended himself.

"A government official? The person you're facing now will never become one."

"Come on, I'll take you to Father, and you can tell him that yourself."

"Not only tell him, with or without you, but I'm quite able even to leave behind this whole family. And you! You touch my things, violating my rights, and don't know you should apologize. Have you never been to school? Or have you indeed never been taught civilized behavior?"

"Shut up! If I'd never been to school, I'd have already ordered you to crawl and make obeisance to me."

"Only a buffalo-brain would think that way about me. Illiterate."

And Mother entered, intervening:

"You meet for the first time in two years . . . why do you have to carry on like village children?"

"I'll fight anyone at all who violates my individual rights, Mother, let alone just a brother."

"Mother, he has admitted all his evil doing in his diary. I was going to present it to Father. He was afraid and went amok."

"You're not yet an official with the right to sell your brother just to obtain some praise," said Mother. "It's not certain that you are any better than he."

I picked up my things.

"It's better I return to Surabaya, Mother."

"No! You have received a task from your father."

"He can do it," I said, looking at my brother.

129

"Your brother is not an H.B.S. student."

"If I am needed, why am I treated like this?"

Mother ordered my brother to another room. After he had left, she resumed.

"You're indeed no longer Javanese. Educated by the Dutch, you've become Dutch, a brown Dutchman, acting this way. Perhaps you've become a Christian."

"Ah, Mother, don't go on so. I'm still the same son as before."

"My son of the past wasn't a rebel like this."

"Your son didn't know right and wrong then. I only rebel against that which is wrong, Mother."

"That is the sign you're no longer Javanese, not paying heed to those older, those with greater right to your respect, those who have more power."

"Mother, don't punish me this way. I respect what is closest to what is right."

"Javanese bow down in submission to those older, more powerful; this is a way to achieve nobility of character. People must have the courage to surrender, Gus. Perhaps you no longer know that song either?"

"I still remember, Mother. I still read the Javanese books. But those are the misguided songs of misguided Javanese. Those who have the courage to surrender are stamped and trodden upon, Mother."

"Gus!"

"Mother, I've studied at Dutch schools for over ten years now in order to find out all this. Is it proper that Mother punish me now I've found out?"

"You've mixed too much with the Dutch. So now you don't like to mix with your own people, even your own family, not even with your father. You won't answer our letters. Perhaps you don't even like me anymore."

"Oh, forgive me, Mother." Her words had struck me sharply. I dropped to the ground, kneeling before her and embraced her legs. "Don't speak like that, Mother. Don't punish me more than my errors deserve. I only know of what Javanese are ignorant, because such knowledge belongs to the Europeans, and because I indeed have learned from them."

She twisted my ear, then knelt down, whispering:

"Mother doesn't punish you. You've discovered your own way. I will not obstruct you, and will not call you back. Travel along the road you hold to be best. But don't hurt your parents, and those you think don't know everything that you know."

"I've never intended to hurt anyone, Mother."

"Ah, Gus, this is perhaps the fate of a woman. She suffers pain when giving birth, then suffers pain again because of her child's actions."

"Please! For Mother to feel pain because of my actions is excessive. Didn't Mother always tell me to study hard and well? I've done that to the full. Now Mother finds fault with me." And as if I were still a little child she caressed my hair and cheeks.

"When I was pregnant with you, I dreamed that someone I didn't know came and gave me a dagger. Since then I've known, Gus, the child in my womb held a sharp weapon. Be careful in using it, Gus. Don't you yourself become its victim.

Since early morning people had been preparing the place for the reception to celebrate my father's appointment. The news was that the best and most beautiful dancers in all the region had been hired for the occasion. Father had brought the best gamelan pure bronze orchestra from T——, my grandmother's gamelan, which was always wrapped in red velvet when not being used. Every year it was not only tuned, but bathed in flower water.

With the gamelan came an expert tuner. My father wanted not only the gamelan itself, but the harmony too, to be pure East Javanese. So since morning, the pavilion had been buzzing with the sounds of people filing things into tune.

The administrative work of the bupati's office stopped altogether. Everyone was helping Mr. Niccolo Moreno, a well-known decorator brought in from Surabaya. He brought with him a big chest of decorating tools the like of which I had never seen before. And it was then that I first realized that arranging decorations and ornamentations was a skill all of its own. Mr. Niccolo Moreno came on the recommendation of Mr. Assistant Resident B——, approved of and guaranteed by Mr. Resident Surabaya.

That morning I too had to meet him. With his own hands he took my measurements, as if he wanted to make me some clothes. Then he let me go.

He had turned the pavilion into an arena whose focus was the portrait of Queen Wilhelmina, that beautiful maiden I had once dreamed after—brought from Surabaya, the work of a German artist named Hüssenfeld. I still admired her beauty.

The Dutch tricolors were hung everywhere, singly or in twos. Tricolor ribbon also streamed out from the portrait to all parts of the pavilion, and would later captivate the audience with its authority. The pavilion's columns were painted with some new kind of paint, made from flour, that dried within two hours. Banyan-tree leaves and greenish-yellow coconut fronds in traditional color harmonies transformed the dry, barren walls into something refreshing, and impelled people to enjoy their beauty. Eyes were drawn by the play of flowers' colors; yellow, blue, red, white, and purple—a saturating beauty—flowers that in day-to-day life stuck separately and silently out along fences.

The big night in my father's life arrived. The gamelan had already been rumbling softly and slowly for some time. Mr. Niccolo Moreno was busy in my room, dressing me up and adorning me. Who would have ever guessed that I, already an adult, would be dressed up by somebody else? A white person too! As if I were a maiden about to ascend the wedding throne.

All the time he was dressing me, he spoke in a strange-sounding, monotone Dutch, as if it came out of the chest of a Native. He obviously wasn't Dutch. According to his story, he often dressed and adorned the bupatis, including my father tonight, and the sultans of Sumatra and Borneo. He'd designed many of their clothes, and even now was often summoned by them. He said also that the costumes of the guards of the kings of Java were designed by him.

Silently I listened to his stories, neither affirming nor refuting them, although I didn't believe them fully either.

He had dressed me in an embroidered vest, stiff, as if made from tortoiseshell. I could never have bent over in it. The stiff leather collar dissuaded my neck from turning around. Indeed the intent was that my body should be straight and stiff, not turning around frequently, eyes straight ahead like a true gentleman. Then a batik sarong with a silver belt. The style in which the batik was worn truly brought out that dashing East Javanese character. That's what Father no doubt wanted. I suffered all this like a young maiden. A batik blangkon headdress, a mixture of East

132

Javanese and Madurese styles, something entirely new, Niccolo Moreno's own creation, was placed upon my head. Then came a ceremonial sheathed short sword, a keris inlaid with jewels. Then a black outer upper garment like a coat with a cut at the back so the people could admire the beauty of my keris. A bow tie made my neck, usually active guiding my eyes to their targets, feel as if it were being snared. Hot perspiration began to soak my back and chest.

In the mirror I found myself looking like a victorious knight out of those stories of the legendary eleventh-century prince, Pan-ji. From under my shirt protruded velvet cloth embroidered with gold thread.

I was clearly a descendant of the knights of Java, so I too was a knight of Java. But why was it a non-Javanese who was making me so dashing? And handsome? Why a European? Perhaps an Italian? Already since Amangkurat I in the 1600s, the clothes of the kings of Java had been designed and made by Europeans, said Mr. Niccolo Moreno. I'm sorry, but your people only wore blankets before we came. Below, above, on the head, only a blanket! His words truly hurt.

Whether his story was true or not, in the mirror I did look dashing and handsome. Perhaps people would say later: "a true Javanese costume," forgetting all the European elements in the shirt, collar, tie, and even forgetting the last and velvet made in England.

I considered my clothes and my appearance to be products of mankind's earth at the end of the nineteenth century, the time of the birth of the modern era. And I truly felt that Java and all its people were a not-too-important corner of this earth of mankind. The town of Twente in Holland now wove for the Javanese, and chose the material too. Village-woven cloth was left now only to the villagers. The Javanese were left with only batik-making. And this one body of mine—still the original!

Mr. Moreno went. And I sat down. When I became aware of the sounds of the East Javanese gamelan, which would cradle this evening's atmosphere, I awoke from my reflections, looked in the mirror again and smiled with satisfaction. In accord with custom, I would be Father's and Mother's escort as they entered the reception. My brother would lead the way, while my sisters had no public function. They would be busy out in the back.

The guests had all arrived. Father and Mother came forth. My brother was in front, I behind them. As soon as we entered the reception area in the pavilion the assistant resident of B—— came up, because that was the program.

All stood in respect. Mr. Assistant Resident walked straight to Father, offered his respects, bowed to Mother, shook hands with my brother and me. Only then did he sit beside Father. The gamelan played a song of welcome, flaring up and filling the reception area and people's hearts. And the pavilion was packed with people, their faces shining with pleasure and the light of the gas lamps. Behind them in the compound, on woven mats, sat rows of village heads and village officials.

The master of ceremonies, the bupati's chief executive assistant, the patih of B——, opened the program. After a moment's hesitation, the gamelan became silent, as if controlled by some supernatural power.

The Dutch national anthem, "Wilhelmus," was sung. People stood. Very few joined in singing. Most, of course, couldn't, only one or two Natives. The others just stood gazing, perhaps swearing at that strange and aggravating melody.

Mr. Assistant Resident B——, as the representative of Mr. Resident Surabaya, began to speak. Mr. Controller Willem Ende came forward, ready to interpret in Javanese. Mr. Assistant Resident shook his head and waved his hand to prevent it. He indicated that I should be interpreter.

For a moment I was nervous, but in a second I regained my character. No, they are no better than you! And that voice gave me courage. Carry out this task in the same way as you take on your exams!

I came to the front, forgetting to bow and stand with my hands clasped before me, according to Javanese custom. I felt as if in front of class. Wherever my eyes wandered they collided with the eyes of the bupatis. Perhaps they were admiring this Javanese knight in his half-Javanese, half-European clothes. Or perhaps they were indulging their antipathy towards me because of my not showing respect towards them.

Mr. Assistant Resident finished his speech, and I finished putting it into Javanese. He shook hands with Father. And now it was Father's turn to speak. He didn't know Dutch, but that was still better than the other bupatis, who were illiterate. He spoke in

Javanese and I put it into Dutch. Now I delivered it in a totally European manner directed at Mr. Assistant Resident B——and the Europeans in attendance. I saw Mr. Assistant Resident nodding, and observing me as if it were I who was giving the speech, or perhaps he was enjoying my act as a monkey in the middle of a crowd. Father's speech ended and so too did my translation. The senior officials stood up and congratulated Father, Mother, my brother, and me.

When Mr. Assistant Resident congratulated me he felt he had to praise my Dutch.

"Very good," then in Malay, "Tuan Bupati, Tuan must indeed be happy to have such a son as this. Not only his Dutch, but more importantly his attitude." Then he resumed in Dutch. "You are an H.B.S. student, yes? Can you come to my house tomorrow afternoon at five o'clock?"

"With pleasure, sir."

"You will be picked up in a carriage."

The congratulations did not take long. The village dignitaries didn't normally shake hands with the bupati. So Father's hand was saved from the twelve hundred or so hands of the village officials. They stayed seated on their mats out in the compound.

The gamelan resumed its tumultuous din. A full-bodied dancer entered the arena as if flying, carrying a tray, upon which was a sash. Carrying the silver tray, she made her way directly to the assistant resident. When the white official rose from his chair, she took the sash and draped it over his shoulder.

People cheered and clapped in approval. He nodded to Father, asking permission to open the tayub dance. Then the assistant resident nodded to the crowd. Unhesitatingly he stepped forward, partnered by the dancer, and moved into the center of the gathering to the crowd's applause and cheering. And he danced, his two fingers holding the corners of the sash, and at every beat of the gong he jerked his head in rhythm with the gong. And before him that full-bodied, pretty, eye-catching woman danced provocatively.

A few minutes later another dancer entered running, also gloriously pretty. With a silver tray in her hand, she entered the arena carrying liquor in a crystal glass. She took up a position beside the assistant resident and joined in dancing.

The official stopped dancing and stood up straight in front of the new dancer. He took the crystal glass and swallowed down three quarters of its contents. The glass with the remaining liquid he pressed to the lips of his dance partner, who drained it down only after trying to resist while still dancing. Then she bowed down her head in extreme embarrassment.

The gathering cheered in glee. The village chiefs and officials stood and contributed to the hubbub.

"Drink it, sweetie! Drink, *hoséééééé!*"

That handsome dancer with her bare, firm, shining, langsat-fruit skin took the glass from the official's hands and placed it on the silver tray.

Mr. Assistant Resident nodded with pleasure, clapping gleefully, and laughed. Then he returned to his chair.

Now another dancer came and offered the sash to Father. And he danced with her beautifully. And that dance too ended with liquor from a silver tray.

Following this, the assistant resident went home. The bupatis too then went home, one by one, each in his own grand carriage. The village chiefs, district officers, police constables, charged the pavilion, and the tayub dance continued until morning with the shout of *hoséééé* after every swallow of liquor.

I only found out the next morning that in my bag there was a small bundle of silver coins. Wrapped in paper, with Annelies's writing: "Don't let us go for long without hearing news from you. Annelies."

The money totalled fifteen guilders, enough for a village family to live for ten months, even twenty months if their daily budget was kept at two and a half cents a day.

That morning I set off to the post office. The postmaster, I don't know his name, an Indo, shook my hand and praised my Dutch at the previous nights' reception as being excellent and very exact. All the office employees stopped working just to listen to our conversation and to take in what I looked like.

"We would be very proud if you would work here; you are an H.B.S. student, yes?"

"I only want to send a telegram," I answered.

"There's no bad news, I hope?"

"No."

The postmaster attended to me himself and gave me the form.

He invited me to sit at a table, and I began to write, then handed the form back to him. Once again he attended to it himself.

"If you have the opportunity, perhaps we could invite you to dinner?"

It appeared that the assistant resident's invitation had become big news in B——. It could be predicted that all the officials, white and brown, would be sending me invitations. So, all of a sudden, I'd become a prince without a principality. How tremendous, an H.B.S. student! in his last year! in the middle of an illiterate society. They will all be out to indulge me. If the assistant resident has started inviting you, naturally you are without flaw, everything you do is right, there is nothing you would ever do that could be said to have violated Javanese custom.

That prediction didn't have long to wait before it was verified. As I left that small office I surveyed the whole area. Everyone bowed in respect. Perhaps among them there were some already sizing me up for their son-in-law or a brother-in-law. "Imagine: an H.B.S. student." And it was true. On arrival back at home I found several letters had already arrived, all written in Javanese script, all invitations!

I didn't know one of the people inviting me. My guess remained the same: They were all thinking of themselves as my future parents or brothers-in-law. "Imagine: the son of a bupati, himself considered to be a future bupati, an H.B.S. student, final year. As young as that and he's already been noticed by an assistant resident. He even defeated the controller!"

B——! A gray corner on this earth of mankind! I spent the whole morning writing apologetic replies: I am unable to accept your invitation because I must return quickly to Surabaya.

And that afternoon the promised carriage arrived to pick me up. I wore European clothes, as was my usual way in Surabaya, even though Mother didn't approve.

It seemed that the news about my invitation had spread throughout all B——. People needed to see me cover that short distance between the bupati building and the assistant-residency building. Unfamiliar people in neat Javanese dress, but with naked feet, bowed in respect. Those wearing hats over their blangkon headdresses needed to raise them.

The carriage took me straight to the back of the assistant residency building, stopping at the veranda.

The assistant resident rose from his garden chair, as too did the two young women beside him. He got in his greeting first.

"This is my eldest daughter," he introduced her, "Sarah. This is my youngest daughter, Miriam. Both are H.B.S. graduates. The youngest went to the same school as you, before you, though, of course. Well, excuse, me, I have some unexpected work to do," and he went.

So this was what the honored invitation rocking B——was all about. I'm introduced to his daughters and then he goes.

Probably Sarah and Miriam were older than I. And every H.B.S. student knew with certainty: Seniors seek every opportunity to put on airs, to strike a pose, to insult and to topple upside-down any poor junior.

You be careful now, Minke. See, Sarah is starting:

"Is Miriam's Dutch language and literature teacher, Mr. Mähler, still teaching? That crazy, talkative one?"

"He's been replaced by Miss Magda Peters," I answered.

"No doubt more talkative still and with only a kitchen vocabulary," she followed on.

"Do you know for sure that she is a Miss?" asked Miriam.

"Everyone calls her Miss."

And Miriam giggled. Then Sarah too. Truly, I didn't know what they were laughing about.

I answered hotheadedly and recklessly:

"I think she has more than just a kitchen vocabulary. She is my cleverest teacher, the one of whom I'm most fond."

Now they both laughed, giggling, while covering their mouths with their handkerchiefs. I was confused, not knowing what was so funny. For a moment I saw shining glances coming from my left and right.

"Fond of a teacher?" teased Miriam. "There has never been a Dutch language and literature teacher whom people have liked. Castor oil dispensers, all of them. What do you get from her?"

"She can cleverly explain the Dutch eighties style and compare it with the contemporary style."

"Oho!" cried Sarah. "If that's the case, try declaiming one of Kloos's poems, so we can see if your teacher really is so great."

"She is clever in explaining the sociological and psychological background of the works of the eighties," I continued.

"Very interesting."

"What do you mean by psychological and social background?"

Sarah and Miriam burst into a fit of giggling again.

Now I was beginning to become annoyed with their giggling. I moved across to the assistant resident's chair to avoid their glances. Now I faced them directly. And they came over as Pure-Blooded girls who were adroit and not at all unattractive. Yet a junior could never relax his vigilance with seniors.

"If you do indeed require an explanation of that," I continued, putting on a serious countenance, "we would need to look at actual literary texts."

Seeing me get more and more into a corner, their giggling escalated and they glanced at each other knowingly.

"Come on, when has there been a Dutch language and literature teacher who talked about social and psychological background? It sounds a lot of hot air to me! What does she want to become, this Miss Magda Peters? At the most she'd be able to present the Dutch Eighties Generation writers who barked at the sky destroyed by the factory smoke, the fields blasted by the din of traffic, under assault by roads and railway lines." Miriam, who was more aggressive, attacked. "If she wants to discuss social background she shouldn't be talking about that sentimental generation, she should be talking about the writer Multatuli . . . and the Indies!"

"Yes, that's when you're really talking about noble literature, where mud has fostered the growth of the water lily."

"She's also spoken about Multatuli," I answered resolutely.

"Ah, come on, how could Multatuli be discussed in school? Stick to the truth. He has never been mentioned in any textbook." Miriam continued her attack.

"Miriam's right," Sarah confirmed. "If one wants to talk about social background, Multatuli is indeed a typical example." Then she glanced at her sister.

"Miss Magda Peters not only put Multatuli forward as a typical example. She went so far as to elucidate his writings."

"Elucidate them!" cried Sarah disbelievingly. "An H.B.S. teacher in the Indies elucidating Multatuli! Could that happen in the next ten years, Miriam?" Miriam shook her head in disbelief. "Or have you changed your textbooks?"

"No."

139

"Your teacher is truly puffed up. You're only her pupil," Sarah tormented me.

"No."

"Then your teacher is really daring. If what you say is true, she could get into trouble." Miriam began to get serious.

"Why?"

"How simple you are. So you don't know. And you need to and indeed are obliged to know." Miriam continued. "Because if what you say about your teacher is true, maybe she is from the radical group."

"There's nothing wrong with the radicals, is there? They're bringing progress to the Indies." By this time I felt really stupid.

"But good doesn't necessarily mean right, and progress might not yet be appropriate. It could come at the wrong time and place!" pressed Miriam.

Sarah cleared her throat. She didn't speak.

"Come on, tell us which of his writings she is enthusiastic about?"

They were becoming more and more annoying. And a junior, I don't know who started the rule, must always show respect. So:

"The main work is of course *Max Havelaar or De Koffie-veilingen der Nederlandsche Handelsmaatschappÿ—The Coffee Auctions of the Netherlands Trading Company.*"

"And who do you think Multatuli is?" Now Sarah was launching an assault on me.

"Who? Eduard Douwes Dekker."

"Excellent. You must also know of the other Douwes Dekker. That's obligatory." Sarah continued her attack.

This mad senior was getting worse and worse. And why was she attacking me like this, glancing at her sister too, lips trembling as she held back her laughter? They're playing out a drama, playing around with a Native slave. They were going too far. There was only one Douwes Dekker known to history.

"So you don't know," Sarah said insultingly. "Or you're in doubt?"

Miriam burst into a fit of uncontrolled giggling.

Very well, I would confront this satanic conspiracy. So this was the value of the honored and sensational invitation from the assistant resident. Very well, because I didn't know, I answered as reasonably as possible.

"I only know of Eduard Douwes Dekker, whose pen name is Multatuli. If there is any other Douwes Dekker I truly don't know of him."

"Indeed, there is," Sarah resumed again. Miriam hid her face in a silk handkerchief. "But more importantly, who is he? Don't be confused, don't go pale," she teased. "You know, don't you, you're just pretending not to know."

"I truly don't know," I answered impatiently.

"Then your teacher Miss Magda Peters, whom you so greatly praise, has insufficient general knowledge. Listen, and remember not to shame your seniors. Don't forget this. The other Douwes Dekker, who is more important than Multatuli, is a youth . . ."

"He's still a youth?"

"Of course he's still a youth. He's on board a ship. Or perhaps he's already in South Africa, fighting with the Dutch against the British. Have you heard of him?"

"No. What has he written?" I asked humbly.

"He's still a youth. So he can, of course, be forgiven if he hasn't yet written anything," answered Sarah, then she too giggled.

"So why should I know about him?" I protested. "People become known because of their works." Now I was getting the chance to defend myself. "Hundreds of millions of people on this earth have not produced works that would have made them famous, so they are not famous."

"Actually he's produced a lot of writings too. But there's only one reader. Here she is, that most faithful of all readers: Miriam de la Croix. He is her boyfriend, understand?"

I swore in my heart. What have I got to do with him if he's only Miriam's boyfriend? These two wouldn't know who Annelies Mellema was either. I'd bet on it!

"Come on, Mir, tell us about your boyfriend," coaxed Sarah in high spirits.

"No." Miriam refused. "It's not anything to do with our guest. Let's talk about something else. You're a pure Native, aren't you Minke?" I was silent, not answering. I felt that, without knocking, they were about to open the door of humiliation. "A Native who has obtained European education. Very good. And you already know so much about Europe. Perhaps you don't know as much about your own country. Perhaps. True? I'm not wrong, am I?"

The humiliation has now begun, I thought.

"Your ancestors," Miriam de la Croix continued,"—I'm sorry, it's not my intention to insult anyone—your ancestors, generation after generation, have believed that thunder is the explosion caused by the angels trying to capture the devil. It's so, yes? Why are you silent? Are you ashamed of your own ancestors' beliefs?"

Sarah de la Croix had stopped laughing. She put on a serious face, and observed me as if I were some mysterious animal.

"There's no need to single out my ancestors," I answered her. "Your European and Dutch ancestors in prehistoric times were no less ignorant."

"Ah," Sarah intervened, "just as I suspected. You two are going to fight about your ancestors."

"Yes, we're like cattle, Minke," Miriam continued. "Fight at our first meeting, but be friends afterwards, perhaps forever. That's right, yes?"

A very adroit girl! My suspicions began to subside.

"My ancestors may have been more stupid than your ancestors, Minke. Your ancestors were building paddy fields and irrigation systems when mine were still living in caves. But that's not what we want to discuss. Look, at school you're taught that thunder is only the clash of positive and negative clouds. Benjamin Franklin is now even able to build a lightning rod. Yes? While your ancestors have a beautiful legend—the story that I have heard—about Ki Ageng Sela, who was able to capture the thunder and then lock it up in a chicken coop."

Sarah burst into laughter. Miriam became even more serious, observing my face as twilight reached its climax. Then she let fly her puzzle:

"I believe you can accept the teachings about positive and negative clouds because you need the marks to pass. But be honest, do you believe in the truth of this explanation?"

Now I knew that she was testing my inner character. Yes, a real test. To be frank, I'd never asked myself such a question. Everything had just seemed to flow smoothly, requiring no questioning.

Now Sarah interfered:

"Of course I believe that you know and have mastered this natural science lesson. But now the problem is: Do you believe it or not?"

"I must believe it," I answered.

"Must believe it only because you've got to pass the exams. Must! So you don't yet believe."

"My teacher, Miss Magda Peters . . ."

"Magda Peters again," cut in Sarah.

"She is my teacher. According to her, everything comes from being taught," I answered, "and from practice. Even beliefs. You two would not ever have come to believe in Jesus Christ without being taught and then practicing to believe."

"Yes, yes, perhaps your teacher is right." Sarah was confused.

Miriam, on the other hand, watched me as if she were looking at her lover's portrait. "This year we've begun hearing a new word: *modern*. Do you know what it means?" The aggressive Miriam began again, forgetting all about the question of thunder.

"I know. But only from Miss Magda Peters."

"It seems you don't have any other teachers," interrupted Sarah.

"What's to be done? It's she who can answer your question."

"Then what does this fantastic teacher of yours mean by *modern*?" Miriam cut in.

"It isn't in the dictionary. But according to this fantastic teacher of mine it is the name for a spirit, an attitude, a way of looking at things that emphasizes the qualities of scholarship, aesthetics, and efficiency. I don't know any other explanation. She is a member of the schismatic group in the Catholic Church that's been expelled by the Pope. Perhaps there's another explanation?" I asked finally.

Sarah and Miriam stared at each other. I couldn't see their faces clearly. Twilight had arrived, though it had seemed ages in coming. And now they just sat silently, exchanging looks, and they began busily to eliminate mosquitoes getting overfriendly with their skin.

"These mosquitoes!" Sarah frowned. "They think I'm a restaurant."

Now it was I who burst into laughter.

"Ah, we've forgotten our drinks," said Sarah. "Please!"

And the tension began to subside. I began to breathe freely again. And I remembered the servant dressed in white who had put the glasses and cakes on the garden table a while ago. For the first time I smiled to myself. Not only because the tension was subsiding, but because I knew that they knew no more than I did.

"Do you know who Dr. Snouck Hurgronje is?" Once again Miriam attacked.

If the assistant resident turned up now, I would be saved from this torment. Where are you, my savior? Why don't you appear? And these children of yours are no less fierce than the twilight mosquitoes. Or did you deliberately invite me here so that your daughters, my seniors, could do me in? These thoughts suddenly made me understand: The assistant resident was deliberately confronting me with his two daughters as a test. He probably had some specific purpose in mind.

"How about if I have a turn asking a question now?"

Sarah and Miriam burst into laughter again.

"Just a minute," forbade Miriam. "Answer first. Your beloved teacher is indeed extraordinary. You, her student, are no less extraordinary. It's only natural you're so fond of her. Perhaps I also would be as fond of her as you are. Now, about my last question, perhaps your beloved teacher has spoken about him too."

"A pity, but no," I answered briefly. "Tell me."

It appeared she had long awaited this opportunity to come forward as a teacher. Skillfully, she told this story:

Dr. Snouck Hurgronje was a jewel of a scholar—daring to think, daring to act, daring to risk himself for the advancement of knowledge—and was an important adviser in ensuring a Dutch victory in the Aceh war. It's a pity he was now involved in an argument with General Van Heutz. An argument about Aceh. What's the meaning of this argument? There isn't any, said Miriam. The important thing is that he has undertaken a valuable experiment with three Native youths. The purpose: to find out if Natives are able truly to understand and bring to life within themselves European learning and science. The three students are going to a European school. He interviews them every week to try to find out if there is any change in their inner character and whether they are able to absorb it all, whether their scientific knowledge and learning from school is only a thin, dry, easily shattered coating on the surface, or something that has really taken root. This scholar has not yet come to a decision.

Now it was I who laughed again. These two misses were aping the scholar. And I was the guinea pig caught by them along the side of the road. Incredible! But they might be doing it on their father's orders, which were probably not ill-intended, so I restrained my desire to launch a counterattack. I continued listening to Miriam's story. Not as a junior, nor as a student—but as an observer.

All was still and calm. Sarah did not speak. Then:

"Have you heard about the Association Theory?"

"Miss Miriam, you are now my teacher," I answered quickly, avoiding the question.

"No, not a teacher," she said with a sudden humility. "These days it's only normal that there should be an exchange of views between educated people. Yes, isn't that right? So you've never heard about it?"

"Not yet."

"Very well. This theory comes from that scholar. A new theory. His idea is that if the experiment succeeds, the Netherlands Indies government could put his theory into practice. That's right, isn't it, Sarah?"

"Tell it yourself," said Sarah, avoiding the question.

"Association means direct cooperation, based on European ways, between European officials and educated Natives. Those of you who have advanced would be invited to join together with us in governing the Indies. So the responsibility would no longer be the burden of the white race alone. So there would no longer be a need for the position of the controller as a liaison between Native and white administrations. The bupatis could cooperate directly with the white government. Do you understand?"

"Keep going," I said.

"What's your opinion?"

"Very simple," I answered. "We Natives have read what you have not read: our chronicle *Babad Tanah Jawi*. Reading and writing Javanese has long been something our families have studied. Look, in E.L.S. and H.B.S. we are taught to admire the Indies Army's brilliance in subjugating us, the Natives."

"The Indies Army is indeed outstanding. That's a fact." Miriam defended her nation.

145

"Yes, indeed, it's a fact. Do you know that the chronicles written by the Natives tell of how we withstood your attacks for centuries?"

"But were always defeated?" charged Miriam.

"Yes, indeed, always defeated." Suddenly the courage to continue my words disappeared. Instead I came out with a question:

"Why didn't you come up with this theory three centuries ago? When no Native would have had any objection to Europeans sharing responsibility with them?"

"I don't quite understand what you mean?" interrupted Sarah.

"I mean, this fantastic scholar, Doctor . . . what's his name again? . . . he's three hundred years behind the Natives of that time," I answered proudly.

And with that I excused myself, leaving those two annoying seniors sitting there.

8

Father and Mother were very proud that I had received an invitation from Assistant Resident Herbert de la Croix.

Invitations from local Native notables continued to arrive at our house.

It was better that my parents didn't know how their son, of whom they were so proud, was made to look a fool.

With all my might I resisted their demands that I tell them what had happened. Instead, I announced that I would quickly be returning to Surabaya.

I was busy replying to all the invitations.

My father was no longer angry with me. The invitation from the assistant resident absolved me of all my sins.

I had telegraphed Wonokromo, giving them the day and time of my return to Surabaya, and asking that I be met at the station by a carriage.

Father and Mother weren't able, and perhaps also didn't feel it proper, to delay my departure. There were no further accusations relating to the issue of Nyai Ontosoroh. Someone who had received an invitation from the assistant resident was immediately immune; it was impossible for him to have done wrong. On the contrary it was as if a sign had appeared over the gateway to his future proclaiming the certainty of his attainment of a high and important position. But they did insist I take leave of and say goodbye to the European official.

I didn't want to go but left for his house anyway. Once again I had to meet with Sarah and Miriam de la Croix. It turned out that when around their father they were not aggressive, but instead orderly and polite.

"Your school director used to be my school friend," said the official. "When you get back to school, pass on my greetings and respects."

Then he explained that his children wanted to go home to the Netherlands. They had been without a mother for the last ten years. If they went, he would be very lonely. Because of that:

"Write to me often about how you're getting on. I will be very happy to read your letters. And correspond with Sarah and Miriam too," he requested. "These days people should exchange views, shouldn't they? Who knows, perhaps such discussions could turn out to be the foundation for a better life later on? Especially if you all become important people!"

I promised I would write.

"Minke, if you maintain your present attitude, I mean your European attitude, not a slavish attitude like most Javanese, perhaps one day you will be an important person. You can become a leader, a pioneer, an example to your race. You, as an educated person, surely understand that your people have fallen very low, humiliatingly low. Europeans can no longer do anything to help. The Natives themselves must begin to do something."

His words hurt. Yes, every time the essence of Java was insulted, offended by outsiders, my feelings were also hurt. I felt so totally Javanese. But when the ignorance and stupidity of Java was mentioned, I felt European. So these messages, which had brought me so many thoughts, I took with me in my heart on the train back to Surabaya.

If Mr. de la Croix had been Javanese, it would have been easy to guess his intention: He wanted me as a son-in-law. But he was a European, so that was impossible, especially as both Sarah and Miriam were several years older than I. The colonial official hoped

I would become a leader, a pioneer, an example to my people. Like in a fairy tale! Nothing like this was ever mentioned in the tales of my ancestors. Is it possible that there could be a European who truly desired such a thing? In the history of the Indies no such thing had ever occurred. The Dutch Army had never rested its rifles and cannons for a single moment during their three hundred years in the Indies. Suddenly there was a European who wanted me to become a leader, a pioneer, and an example to my people. An uninteresting fairy story. An unfunny joke. It appeared he wanted to make me a guinea pig in an experiment to test the Association Theory of Dr. Snouck Hurgronje. To the devil with it all! Nothing to do with me. It's lucky I liked to make notes, so I had a treasury that at every moment could supply me with guidelines and warnings.

I groped in my suitcase to find and read those letters that I had still not yet read. It was true, they told of the plan for a reception for the investiture of Father, and ordered and requested that I come home quickly. With my brother's letter there was even a note to my school director requesting leave.

Suddenly, on the train, I noticed a fat man with rather slanted eyes, watching only me. His clothes were made from brown drill, both his shirt and his trousers. His shoes were brown too—shoes that were normal for first-class carriages. His hat, made from felt, with a silk hatband, never left his head, but was sometimes lowered to cover his forehead in order to give him the freedom to look around the carriage as he liked. His baggage consisted of a small leather case that had been placed on the rack over his head. And he sat on the bench over there, on the other side. When the conductor was checking the tickets, Fatso handed over his white ticket, but his eyes glanced across at me.

Between B—— and Surabaya there were only a few places where the express train stopped. And Fatso didn't get off, nor were there any signs he was getting ready to do so. He was obviously heading for the last station. "Stop!" I thought to myself. I'm not going to take any notice of him. I want to enjoy this trip, as if it were a holiday. I wanted to sleep deeply. I needed my strength and health.

The train hissed on rapidly towards Surabaya. By five in the evening Surabaya rattled under our wheels. The train shot past the cemetery and stopped at the station. The platform looked deserted. Just a few people were standing or sitting, waiting or walking back and forth.

"Ann! Annelies!" I called from my window. She was there to meet me.

Annelies ran to my carriage, stopped below it, and put out her hand.

"Everything's all right, Mas?" she asked.

Fatso passed me carrying his little case. He alighted first, looked briefly at Annelies, then slowly walked towards the station exit. I followed him with my eyes. He didn't go out, but stopped and glanced back at us.

"Come on, get out. What else are you waiting for?" coaxed Annelies.

I alighted. The coolie followed, carrying my baggage.

"Come on, Darsam has been waiting for a long time."

Fatso still hadn't gone out through the platform gate, so we ended up passing by him. His skin was clear, the color of langsat fruit, his face reddish. Every few moments he wiped his neck with a blue handkerchief, just as he had done in the carriage. As soon as we passed him, he moved as if he wanted to follow us.

"Greetings, Young Master!" called Darsam from beside the cart. (Mama had told him he was not to call me Sinyo.)

Fatso watched us as we climbed up into the carriage. Now I really began to be suspicious. Who was he? Why hadn't he gone yet, why was he still watching us? And as soon as we boarded, he hurriedly rented a carriage. As soon as our cart moved, his carriage set off. It was obvious he was following us.

When I glanced back at his carriage, he was wiping his neck. He was not paying any attention to us. The second time I looked, he was looking at us.

"Hey! Darsam! Why aren't you turning right?" I protested.

"Why to the left, Darsam?" asked Annelies in Madurese.

"I've got a little business," he answered briefly.

The carriage turned left away from the station square, then to the right, passing the green field in front of the residency building. And where was Darsam off to? And he looked so serious.

"Why aren't you turning right again?" protested Annelies. "It's already evening."

"Patience, Non, it's not dark yet. A lantern has been prepared. Don't worry."

And it was turning out that Fatso's carriage was indeed following our cart. When, for the umpteenth time, I glanced back he bent down, sheltering his face behind the driver's back.

"Slow down a little, Darsam," I ordered.

The cart entered into a low-class street, traveling slowly. The carriage behind us also slowed down. It had to—the street was too narrow. The carriage could have rung its bell if it wanted us to move over. But it didn't. Nor did it try to overtake us.

All of a sudden our cart stopped.

"Why here?" protested Annelies.

"Just a moment, Non. I have a little business," answered Darsam, jumping down and guiding the horse to the side of the road, tying the reins to a fence post.

Fatso's carriage hesitated to pass, but in the end it had no choice. The passenger himself turned the other way while blowing his nose with his blue handkerchief. He didn't look Chinese, or like a Mixed-Blood Chinese, nor like a merchant. Anyway, if he was a Mixed-Blood Chinese, he was probably an educated one, perhaps an employee at the office of the Majoor der Chinezen—the Dutch-installed leader of the local Chinese community? Or perhaps a Mixed-Blood European-Chinese returning from holidays to his workplace in Surabaya? He was clearly not a merchant. They weren't the clothes of a trader. Or perhaps he was a cashier at one of the "Big Five" Dutch trading companies—Borsumij or Geowehrij? Or perhaps he was the Majoor der Chinezen himself? But the majoors were always arrogant, considering themselves equals with Europeans, and so wouldn't take any notice of me, or any other Native for that matter. Or perhaps he was interested in Annelies? No. This had been going on since I left B——.

"Non, wait here a moment. Darsam has a little business in this food stall," said Darsam. With his eyes directed at me, he continued, "Young Master, could you come down for a moment?"

I climbed down. Vigilantly, of course. We entered the café, a bamboo shack with a tiled roof.

"What's going on there?" asked Annelies suspiciously, from the top of the cart.

Darsam glanced back, answering:

"Since when hasn't Noni trusted Darsam?"

I was also becoming suspicious. Fatso and his carriage had

stopped some way down the road. Now it was Darsam who was up to something.

"Stay there, Ann," I said to calm her. Yet I felt my eyes following the hands and machete of the Madurese fighter.

There was only one customer drinking coffee in the stall. He didn't look up when we entered. It looked as if he was daydreaming. Or pretending not to notice? Or an ally too of Fatso, like Darsam perhaps?

In the manner of giving an order he invited me to take a seat on the long bench across from the other customer. He sat so close to me that I could hear his breathing and smell his sweat.

"Take some tea and cake to the carriage outside," Darsam instructed the stall woman. His eyes scrutinized her sharply until she took the food out on a wooden tray.

His eyes shone wildly as he brought his curly mustache up close to me, and whispered in heavy and awkward Javanese:

"Young Master, something has happened at the house. Only I know. Noni and Nyai don't. Young Master mustn't be startled. For the moment Young Master mustn't stay at Wonokromo. It's dangerous."

"What's the matter, Darsam?"

Now his voice was calmer:

"Darsam is loyal only to Nyai, Young Master. Whoever is loved by Nyai is loved by Darsam. Whatever she orders, Darsam carries out. I don't care what sort of order it is. Nyai has ordered me to look after Young Master, so I will do it. Young Master's safety is now my work. You don't need to believe everything I say, Young Master, but at least take my advice."

"I understand your task. Thank you for being so conscientious. But what has happened?"

"Nyai is my employer. Noni is my employer too, but only number two. Now Noni is in love with Young Master. Darsam must also make sure that nothing happens to you. So I pass on this advice. Not because Darsam's machete can't guarantee your safety. No, Young Master. There is still something not yet clear to Darsam."

"I understand. But what has happened?"

"In short, Darsam will take Young Master back to his room at Kranggan, not to Wonokromo."

"I must know why."

He was silent and closely observed the stall woman as she went by.

"Finished yet, Darsam?" came Annelies's voice.

"Be patient, Non," he answered without looking outside. Seeing that the stall woman had gone, he resumed his whispering. "It is Sinyo Robert, Young Master. Making many promises, he has ordered me, Darsam, to kill Young Master."

I wasn't at all surprised. I had already noted the signs of that youth's ill intent.

"How have I wronged him?"

"Only jealousy, I think. Nyai is fonder of Young Master. He feels unhappy with another man in the house."

"He can tell me to my face. Why is he going to you?"

"He thinks too little, Young Master. That's what makes him so dangerous. Now Young Master knows, understands my advice. Don't tell this to Nyai or Noni. Don't ever. Let's go." He paid for what we'd eaten, not asking my opinion about any of this.

Fatso's buggy had disappeared.

Our carriage set off. And in Wonokromo—if Darsam was right—someone wanted to take my life, the only one I had. Fatso had been spying on me since B——. Perhaps my father's anger with me was justified after all. And my mother's warning that I must be ready to accept all the consequences of my own deeds was not to be wasted either.

Yes, yes, Robert Mellema had the right to look upon me as an intruder into his kingdom. At the very least I was another thing for him to worry about. He was fully entitled to think that way about me.

Annelies didn't want to let go of my hand, as if I were a slippery fish that might jump out of the carriage at any moment. She didn't speak. Her eyes showed her thoughts were far away.

"Ann, I found your money in my case," I said.

"Yes, I put it there. You might have needed it. You were on an unknown journey and you had to return to me quickly."

"Thank you, Ann. I didn't use it."

She laughed for the first time. But her laughter didn't interest me. The carriage lantern didn't throw its rays back into the carriage. Darkness. Annelies's beauty was swallowed up by the blackness. Even if that hadn't been so, I'd still not have been interested. My mind was preoccupied with more forbidding matters, and

153

these matters were stealing all that could be said to be enjoyable.
My earth, this earth of mankind, had lost all its certainty. All the
science and learning that had made me what I am evaporated into
nothingness. Nothing could be trusted. Robert? Yes, I understood
him. Fatso? I would recognize his shape, even, I think, in the dark.
But it could be someone I didn't know, someone I'd never predict,
who was going to carry out this evil against me. Surabaya was
famous for its paid killers—who charged only a half to two rupi-
ahs. Every week there were corpses found sprawled on the beach,
in the forests, on the roadside, in the markets, and their bodies
always had knife wounds.

The carriage headed for Kranggan.

"Why are we going this way?" Annelies protested again.

What could I say to Annelies? before I'd a chance to think up
an explanation, we stopped in front of the Telingas' house. Without speaking Darsam off-loaded my things.

"Why are they being off-loaded here?" protested Annelies
again.

"Ann," I said gently, "I have to prepare for my lessons this
week. So for the time being, I can't accompany you home. I'm
really sorry. Thank you for meeting me, Ann. Say sorry to Mama
for me, yes? I really can't go on to Wonokromo yet. I must stay
here where I am closer to my teachers. My regards and thanks to
Mama. Once I'm free again I'll definitely come back to Wonokromo."

"Mas hasn't been able to study while he's been at Wonokromo? No one bothered you. Forgive me if I've been a bother to
you." She was on the verge of crying.

"No, Ann, of course not."

"Tell me if I've been a bother, so I know what I've done
wrong." Her voice trembled as she came closer to crying.

"No, Ann; truly, no."

There was no way it could have been avoided. She cried.
Cried like a little child.

"Why are you crying? It's only for a week, Ann, only a week.
After that I'll be there again for sure. Isn't that so, Darsam?"

"Yes, Non. Don't cry like this at somebody else's house."

At that moment my sense of being a Javanese knight disappeared; a knight beyond compare, but only in my own imagina-

tion. Now I was just a coward—scared because of a report, just a report, that my life was in danger.

"Don't get down, Ann, just sit here in the carriage," and I kissed her on her cheek in the darkness of the vehicle. I felt wetness on her face.

"Mas must come quickly to Wonokromo," she entreated, crying, surrendering to her feelings.

"So you understand, yes?" She nodded. "When everything is over, I will come back quickly. For the moment I hope that you will listen to me and understand my situation."

"Yes, Mas, I'm not disagreeing," she answered faintly.

"Until we meet again, my goddess."

"Mas."

I got down. Darsam was still waiting in front of the door. It was already night and lamps were shimmering everywhere. It was only my thoughts that weren't clear.

"Why don't you tell Mama?" I whispered to Darsam.

"No. Nyai already has enough troubles because of her children and her tuan. Darsam must take care of this matter himself. Young Master must be patient."

Mr. and Mrs. Telinga were on the settee waiting for me to come out of my room and tell them what had happened. Such a good and happy couple! I don't know how they felt about me. I didn't go out, but locked the door from inside, changed clothes, and got into bed without eating dinner. Before I put out the lamp I still needed to look up at the portrait of Queen Wilhelmina. This earth of mankind! She was secure in her palace, free of all problems, except perhaps within her own heart and mind. And me? Her subject, promised by the astrologer's stars the same fate, yet I may glimpse at any moment, springing from out of the corners of this room, a death arranged by Robert Mellema.

The room was enveloped in darkness. The conversation in the middle room came indistinctly to my ears. No meaning. As young as this, and already there was somebody desiring to take my life. The promise of the modern age, glorious and exciting as told by my teachers, showed not a single sign or indication of its whereabouts. Robert, why are you as mad as this? Murder because of jealous love still occurs the world over—a remnant of bestiality among men. Something unique to man. Murder out of greed was another remnant from man's bestial life. Yes? It's true, isn't it? But you, you're more complex still. You hate your mother, your origins, and obtain no love from them. You beg love from your father, who pays you no heed either. You're jealous, Robert, because your mother's love now flows to me. Because you're afraid that your inheritance will flow to me also—somebody who has no right to it just as depicted in European fiction. Probably in your eyes I'm no better than a criminal.

I've been honest to myself, haven't I? And to the world? Look: I want no more than to enjoy what I've created by my own hard strivings. I need nothing else. A happy life, in my view, does not come from that which is given to one, but from one's own struggles. My separation from my own family had taught me that—a problem more complicated than all my school lessons put together.

And you too, Darsam! Let's hope your mouth cannot be believed. Let's hope Robert is not as evil as that. But you too are hiding some other purpose, and an evil one.

And you, Fatso, with clear, langsat-fruit skin and slightly slanted eyes—what is your business with me? Someone as neatly dressed as you, is it possible you're just a paid killer? Because you want the Mellema girl and the Mellema wealth?

And Sarah and Miriam de la Croix, and Assistant Resident B—— . . . and Association Theory . . .

My heart shriveled up. Why was I such a coward?

9

So that this story of mine runs in order, let me next relate what happened to Robert after I left Wonokromo for B—— in the company of that police officer.

I've put together the story below based on what Annelies, Nyai, Darsam, and others told me; and this is how it has ended up:

When the buggy I was traveling in disappeared, swallowed up by the morning darkness, Annelies cried, embracing Mama. (I don't know why she was such a crybaby and so spoiled, just like a little child.)

"Quiet, Ann, he'll be safe," said Nyai.

"Why did Mama let him be taken away?" protested Annelies.

"That slave of the law, Ann; there's no way we can fight him."

"Let's follow him, Mama."

"No point. It's still too early. And it's clear he's being taken to B——."

"Mama, ah Mama."

"You really love him?"

157

"Don't torment me like this, Mama."

"So what should I do? Nothing, Ann. We must wait. We can't just do whatever we feel like."

"Do something, Mama. Do something."

"You think Minke is just your doll, Ann. He's not a doll. Do something, do something! Of course, I'm going to do something. Be patient. It's still too early in the morning."

"You're going to leave me like this, Mama? Do you want to kill me?"

Nyai became confused. She had never heard that sort of lamentation from her daughter, who usually never complained about anything. She knew and understood that Annelies was going through a crisis. Annelies: her trusted work companion. She knew she must do everything that Annelies wanted, that Annelies considered to be her right. She took her daughter inside to let her rest in her room.

But Annelies refused. She wanted to wait for Minke until he returned.

"It's not possible, Ann. Not possible. Maybe it'll be the day after tomorrow or the next day before he's back."

And Annelies began to withdraw into silence.

Mama became even more confused. She knew that ever since Annelies was little she had never asked for anything. And now over the last few weeks she had been asking—not just asking, but urging, demanding—all concerning Minke. She had always been obedient, behaved sweetly, had been sweet-hearted. Now she was beginning to be rebellious.

Annelies demanded her doll back. And the only person she could go to was her mother.

Nyai worried that her daughter might fall ill. She began to see more and more signs that all was not well with Annelies. Could it be that this obedient child was unable to deal with personal trauma, just like her father?

And the sun slowly began to rise.

Darsam arrived to open the doors and windows. He was startled to see the behavior of his Noni. He was powerless to solve a problem requiring neither machete nor muscles.

"Yes, this is government business," said Mama softly, in a rustling voice. "Affairs that can neither be felt nor seen, affairs of the spirit world."

Suddenly Mama remembered her eldest child. In the next moment, she suspected him of sending an anonymous letter to the police. She suspected him! She would investigate immediately.

"Call Robert here," Mama ordered Darsam.

And Robert came, rubbing his eyes. He stood silently. If it hadn't been for Darsam, he wouldn't have come. Everyone knew that. He stood without speaking a word. His eyes were dull with disinterest.

"How many times and to whom have you sent your poison-pen letters?"

He didn't answer. Darsam went up closer to him.

"Answer, Nyo," the fighter urged.

Annelies was clinging to Nyai, holding herself up.

"I've got nothing to do with any anonymous letters," he answered viciously, his face towards Darsam. "Do I look like a poison-pen letter writer?"

"Answer to Nyai, not to me," hissed Darsam.

"I've never written any, let alone sent any." Now he faced Mama.

"Good. I always try to believe what you say. Why do you hate Minke? Because he's better and more educated than you?"

"I've got no business with Minke. He's only a Native."

"And it's because he's a Native that you hate him."

"So what's the point of having European blood?" he challenged her.

"Good. You hate Minke because he is a Native and you have European blood. Good. It's obvious I'm not capable of educating and teaching you. Only a European could do that for you. Good, Rob. Now I, your mother, this Native, know that people with European blood are, of course, wiser, more educated than Natives. You know what I mean. Now, I ask the Native blood in you—not the European in you—to go to the Surabaya police station. Find out what's happened to Minke. Darsam can't do that. I can't either. The work here won't allow it. You speak Dutch well and you can read and write. Darsam can't. I want to see what you're capable of doing. Go by horse, and be quick."

Robert didn't reply.

"Go, Nyo!" ordered Darsam.

Without answering, Robert Mellema turned around and

walked off, dragging his sandals. He went into his room and didn't come out again.

"Warn him, Darsam!" ordered Mama.

Darsam went after Robert into his room.

The world outside was beginning to clear. Robert left his room escorted by the Madurese fighter. He went out back, to the stables. He had put on jodhpurs and riding boots, and in his hand was a leather whip.

"You just sleep, Ann," consoled Nyai.

"No."

Nyai checked Annelies's temperature and it was rising. The girl was falling ill. Her mother was very anxious.

"Put the sofa in the office, Darsam, so I can be with her while I work. Don't forget the blanket. Then fetch Dr. Martinet." She sat her child on the chair. "Be patient, Ann, patient. Do you really love him?"

"Mama, my Mama!" whispered Annelies.

"Falling sick like this, Ann! No, Mama won't forbid you loving him. No, darling. You can marry him, any time you like, if he agrees. But now, be patient."

"Mama," called Annelies with her eyes closed. "Where's your cheek, Mama; here, Mama, so I can kiss it," and she kissed her mother's cheek.

"But don't fall ill. Who will help me? Could you bear to watch your Mama work like a horse?"

"Mama, I'll always help you."

"So you mustn't get sick like this, darling."

"I don't want to be sick, Mama."

"Your temperature is rising, Ann. You must learn wisdom, child; people can only do their best, then be patient in awaiting the outcome."

Darsam moved the sofa into the office, but Annelies refused to be moved until she'd seen Robert leave on his horse. And her brother still wasn't to be seen.

"Chase Robert, Darsam!" exclaimed Mama.

Darsam ran to the back. Ten minutes later that tall, handsome youth raced off on his horse without looking back, straight down onto the main road. And a quarter of an hour later Darsam drove off in the buggy to fetch Dr. Martinet.

Only then was Annelies willing to be led into the office. Nyai lay a brown-onion-and-vinegar compress on her daughter's forehead.

"Forgive me, Ann, I'm not strong enough to carry you. Sleep now. The doctor will be here in a little while, and Robert will be back with news."

Nyai went to the corner of the office, turned on the tap, and washed her face and combed her hair.

From under the blanket, Annelies asked in a whisper:

"Do you like him, Mama?"

"Of course, Ann, a good boy," answered Nyai, still combing her hair. "How could Mama not like him, when you do? Any parent would be proud to have a son like him. And what woman wouldn't be proud to be his wife one day? His legal wife. Mama too would be proud to have him as a son-in-law."

"Mama, my own Mama!"

"So you shouldn't worry about a thing."

"Does he like me, Mama?"

"What boy wouldn't be mad about you, Ann? Pure-Blood, Indo, Native. All of them. Mama knows, Ann. There's no girl as beautiful as you. Don't worry about a thing. Close your eyes."

The girl's eyes had already been closed for some time. She asked:

"If his parents forbid it, Mama, then what?"

"I told you not to worry about anything. Mama will arrange everything. Sleep. Stay quiet there. Let me get some milk. Remember, you must be healthy. What would Minke say if you became unattractive and gloomy? Even the prettiest girl looks unattractive when she's sick."

Nyai called out from the office to someone from the kitchen. Not long after, someone brought some hot milk.

"Mama will bathe you first. Then try to sleep, Ann."

Nyai went to bathe. On her return she brought warm water and a towel and took care of her daughter.

Annelies didn't say a single word.

Dr. Martinet came, examined her for a moment, and then gave her some medicine. He was fortyish, polite, quiet, and friendly. He was dressed all in white except for his gray felt hat.

161

In his right eye was a monocle attached by a gold chain to the top buttonhole.

Darsam hurried around preparing breakfast for the doctor to eat in the office. And the guest breakfasted with Nyai.

"I'll come back this afternoon, Nyai. Give her some breakfast before she sleeps, but no solids. Keep her away from any noise or commotion. Make sure everything is quiet. Sleep is her best medicine. Move her into her own room. Don't leave her in the office like this. Or move the sofa into the middle room. Keep the windows and the doors closed."

And what about Robert Mellema?

According to the people at Boerderij, eyewitnesses, and the accused at a trial later on, the events unfolded as I've assembled them below:

After leaving the stables, Robert raced the horse along the road. Then he turned right towards Surabaya. When he got onto the main road he pulled up his horse, looked left and right, and slowed down to enjoy the morning view. He probably felt resentful. Just to protect an adventurer like Minke he had to wake as early as this and go to a police station too. And what for? Let Minke disappear forever. The world won't be any poorer, won't have to undergo any extra suffering without him; a speck of dust brought in by the wind from who knows where, and wanting to dwell in his house for who knows how long.

The horse walked on unhappily because, of course, it hadn't eaten that morning, hadn't had its sweet drink yet. Robert hadn't breakfasted yet either, and already he had to be off working.

The morning was more than just cool. The buffalo carts carrying the oil drums from Wonokromo hadn't appeared yet in their usual long, seemingly unbroken convoy. Only the traders from the villages were out walking in a line, carrying on their backs produce for the markets of Surabaya.

The horse had covered about fifty meters at a slow walk. Robert's thoughts were wandering everywhere. From behind the hedge on the right a voice could be heard calling out a greeting.

"Regards, Sinyo Robert."

He pulled up his horse and had a look over the top of the hedge. He could see a Chinese man in striped pajamas smiling sweetly at him. The man had very little hair so that even his pigtail was very thin. When he smiled, his cheeks pulled upward and his eyes became even more narrow and slanted. Even his mustache was thin, long, drooping impotently at the ends of his mouth. His beard was also very thin, and out of a birthmark, a part of his beard formed a tassel and was darker.

"Greetings, Nyo," he repeated, when he saw Robert was unsure whether to answer.

"Greetings, Babah Ah Tjong!" Robert answered politely, nodding and smiling.

"Regards, regards, Nyo. How are things with Nyai?"

"Well, Babah. This is the first time I've seen Babah. Where have you been all this time?"

"As usual, Nyo, much business. And how are things with Tuan?"

"Well, Babah."

"I haven't seen him around for a long time."

"As usual, Bah, a lot of business. The door to Babah's house is open today. The windows too. What's going on today, Bah? Something special perhaps?"

"A good day, Nyo. A day for pleasure. Come on, Nyo, drop in." Ah Tjong's smile mollified Robert's resentment and also his hatred of everything Chinese. He'd never had any desire to meet a Chinese. On any other occasion, he wouldn't even have responded to a greeting from one, let alone enter his yard or house; but now there was something he really wanted to find out.

"Good, Bah, I'll drop in for a moment," and Robert turned his horse into his neighbor's yard.

He had never met Babah Ah Tjong, so he had only guessed that this was who he was speaking to. Babah ran down to greet Robert. Robert saw the pigtailed man clap his hands. A sinkeh, a full-blooded, immigrant Chinese, a gardener, came running, and he took the horse from Robert's hands, then led the horse around to the back of the house.

Robert and Ah Tjong walked together slowly along the rocky path towards that building whose doors and windows were usually never open. They entered. And now the front stairs vanished behind a curtain of coconut-husk cords. The front area, which had no veranda, was very large, furnished with a number of carved teak settees. In one corner there was a brown-spotted bamboo settee. The walls were decorated with different-sized mirrors with red Chinese calligraphy on them. A carved wooden partition closed off the mouth of the corridor in the middle of the building. Several big empty porcelain vases decorated the room; they stood on legs with a dragon curled around them. There were no floor decorations. Neither was there a picture of Queen Wilhelmina. There were no flowers anywhere in the front room either.

Ah Tjong took him to the bamboo settee, which consisted of three chairs and a long bench. The bench faced the front court-yard. The host sat there and Robert opposite him.

"Ah, Nyo, we've been neighbors so long and you've never come to visit."

"How could I when the doors and windows are always closed?"

"Ah, Nyo, come now. How could this house be kept shut all the time?"

"This is the first time I've seen it open."

"If it's opened up like this, Nyo Robert, it means, of course, that I'm home."

"So where do you go when it's closed up?"

"Where do I go?" he laughed happily. "What will you drink, Nyo? What's your usual? Whiskey, brandy, cognac? Chinese wine perhaps? White, yellow, warm, cold? Or Malaga wine? Or dry?"

"Ah, Babah, as early as this."

"What's wrong with that? With fried peanuts, heh?"

"I agree, Bah, agree completely."

"Good, Nyo. It's pleasing to receive a guest like Sinyo: hand-some, dashing, not shy, young . . . Sinyo has everything. Wealthy . . . *wah*."

He clapped hands haughtily, without moving his head, with-out turning, just like a sultan. From behind the partition emerged a Chinese girl in a long, sleeveless gown. The side of her gown was split high up, exposing her thigh. Her hair was braided into two pigtails.

Robert stared wide-eyed at the alabaster-skinned girl. His eyes couldn't move from the split in her gown until the girl came up close to put the whiskey bottle, glasses, and fried peanuts on the table.

Ah Tjong spoke quickly in Chinese to the girl, who then stood up straight in front of Robert.

"Nyo, look this girl over."

Robert was acutely embarrassed. He couldn't speak. His eyes and face shifted away as if tugged by a demon.

"This is Miss Min Hwa. Sinyo doesn't like her?" He cleared his throat. "Just out from Hong Kong."

Min Hwa bowed, put the tray on the table, and sat on the chair near Robert.

"It's a great pity, Nyo. Min Hwa can't speak Malay or Dutch or Javanese. Only Chinese. What can one do? Why is Sinyo silent? Why? She's next to you now. Ai-ai, don't pretend Sinyo has never done this before! Come on, Nyo, you don't need to be shy with Babah."

Min Hwa pressed the whiskey glass to Robert's lips, and he took it hesitantly.

Ah Tjong smiled sweetly, deliberately encouraging him. Min Hwa laughed shrilly and friskily, throwing her head back, her face muscles pulled tight, her mouth open and her pearl teeth, except for one which was gold, on display inside. Then the girl spoke quickly and loudly without pauses, without full stops. Robert didn't understand; instead he became even more unsure of himself as the girl moved her chair closer.

Seeing Robert go pale and the glass in his hand almost fall, Min Hwa pushed the glass up to the tall youth's lips again. And Robert swallowed the whiskey down without hesitating. Suddenly he started coughing—he had never drunk liquor before. Whiskey sprayed all over Ah Tjong and Min Hwa. They weren't angry, but laughed happily.

"Another glass, Nyo," the host suggested.

Min Hwa poured more whiskey into the glass and once more ordered the young guest to drink. He refused and wiped his mouth with a handkerchief. He was even more embarrassed.

"Come on, Sinyo, you aren't going to pretend you've never drunk whiskey before?" he teased. "You don't like whiskey, you don't like Min Hwa?" He waved his hand and the girl left, disappearing behind the carved partition. He clapped again.

Now another Chinese girl appeared, wearing a silk shirt and bright-colored pants. She wiggled as she walked up to the bamboo settee carrying a bamboo tray on which were various delicacies. She put it on the table, on top of the tray left by Min Hwa.

She bowed to Robert and smiled enticingly. Like the first girl, she wore lipstick. Before the delicacies were all laid out Min Hwa entered again, bringing a glass of clear water on a glass tray. She put it before Robert. Then she sat down again on the same seat as before.

"Ah, Nyo, there are two now. Which is the more interesting? Come on, don't be shy. This one is Sie-Sie."

Several carriages began arriving at the front of the house. The guests all came straight inside. Some wore Chinese clothes, others pajamas. All were men and had pigtails. Without worrying whether the host was there or not, they sat down straight away and began busily chatting, laughing, and gambling.

"It looks like there's none you like, Nyo," breathed Ah Tjong and he moved his hand to order them to leave to serve the other guests. "Sinyo doesn't like Sie-Sie either. . . ." He stood and called Sie-Sie over.

As soon as the woman had come back, Ah Tjong sat her down next to Robert.

"Who knows, perhaps Sinyo prefers this one."

And Robert still appeared very embarrassed, confused; wanting to, but afraid. Babah broke into laughter again, enjoying seeing the youth in his confusion. The other guests took no notice of the three of them sitting in the corner.

Sie-Sie started chatting in a loud, fast voice; then she began to seduce him, straightening his shirt and belt, pinching the crease on his shirt. Babah kept observing it all and laughing too. Robert shrank up still more. Then the two Chinese spoke noisily to each other. Robert still couldn't understand a word.

"Very well, Nyo; Nyo doesn't like either of them."

Sie-Sie rose and disappeared behind the partition and Ah Tjong clapped four times.

Robert began to regret her going. He bowed his head. From behind the partition there now appeared a Japanese woman in a kimino with big flowers on it. She took short, quick steps. Her face was reddish and round and her lips were lipsticked and always smiling. Her hair was in a bun. She sat straight down beside the host. When she laughed, one gold tooth was visible.

"Look here, Nyo; here is another one."

Perhaps because he didn't want to have still more regrets, Robert got up the courage to look at the Japanese woman.

"Nyo, this is Maiko. Just two months out from Japan."

Before he stopped talking, Maiko spoke in a high voice in rapid Japanese. Robert didn't understand this either. Yet he got up the courage to gaze at her.

Ah Tjong put his hand across the woman's mouth and said: "This is my own one. Sinyo can have her if you like. Sit here, near her."

Like a dog scared of his master's stick, Robert stood and slowly moved over to sit on the bench, so Maiko was squeezed between them.

"So Sinyo likes this one? Maiko? Good." He laughed, understanding. "In that case I'll go. I'll leave it up to Sinyo."

The guest followed his host with his eyes.

Ah Tjong mixed with his many guests, playing cards, billiards, or mah-jongg. He walked around slowly, checking each table. Then he went back to the bamboo settee and stood in front of the couple, neither of whom could say a meaningful word to the other.

"Yes, it's difficult, Nyo. Maiko doesn't understand Malay, let alone Dutch. How come Sinyo has never mixed with Japanese ladies? You've never been to the Kembang Jepun, perhaps?"

"I've never even seen one before now, Bah." Robert had at last got up the courage to speak.

"A loss, Nyo, a loss for a youth with money. In nearly every Chinese pleasure-house like this, there is a Japanese miss. A loss, Nyo, a loss. You've never been into places in the red light district downtown? In the Kembang Jepun? In Betawi? Indeed you've really missed out on something, Nyo Robert . . . Japanese misses everywhere . . . a pity. Come on."

He summoned Robert with an emperor's flourish, and the three of them left the room, Babah out front, Robert behind him, and Maiko at the rear. Ah Tjong's pigtail swayed a little at each step because it was so thin and it swept across the back of his pajama shirt. They passed the carved partition. Maiko continued to talk in her enticing voice and to walk in those short, quick steps. The smell of perfume filled the air.

They entered a corridor that was hemmed in on left and right by rooms, and that had no furniture except wall decorations. Here and there a few young Chinese girls were standing talking to each other. They were all elaborately dressed and neatly made up and greeted Ah Tjong with great respect, then Robert, but not Maiko.

Robert paid attention to every person. Short, tall, thin, fat, well built and weak-looking; they all wore lipstick and smiled or laughed.

"Such pretty girls are life's pleasure, Nyo. It's a pity you don't like the Chinese ones." He laughed piercingly. "All the rooms face each other. Sinyo can take whichever he likes, as long as the door is left unlocked."

He opened a door so Robert could see inside. Its furniture was as good as that in his own room, and it was just as clean, only it wasn't as big, and the bed was more beautiful.

"For Sinyo here is a king's room, a room of honor, if Sinyo likes it." He walked along again and opened another door. "Only Tuan Majoor may use this room. It happens he's in Hong Kong at the moment."

All the furniture inside was new and in a style Robert didn't recognize—had never come across. Babah asked his guest's opinion. And Robert had no opinion except to agree that it was beautiful.

Ah Tjong entered; Robert and Maiko followed.

"The best furniture, Nyo. Just finished, French style in teak. Made by famous French craftsmen. Indeed Tuan Majoor likes everything French. Most expensive furnishings in the building, Nyo. In the corner is a little cupboard, on top of that little table there's whiskey and sake, whatever Sinyo likes. Settee, sofa, and divan," he said, pointing out each in turn. "Such a beautifully carved wooden bed makes for restful and pleasant sleeping. Yes, Maiko?"

And Maiko answered with a bow and with a soft, fast voice, like a magpie.

"Nyo, enjoy yourself!"

Robert's eyes followed Ah Tjong as he walked outside, watching his pigtail until it disappeared behind the door.

10

Because I consider the time sequence to be important I've also prepared this section from material I obtained from the court testimony later. Most of it is based on Maiko's—given through a sworn translator and written down by me just as I heard it but in my own words.

I went to Hong Kong from Nagoya, Japan, where I was born. I went as a prostitute. My boss was a Japanese and he then sold me to a Chinese boss in Hong Kong. I can no longer remember the name of my second boss. Just a few weeks in his hands were not enough for me to be able to remember his name, so hard to pronounce. He sold me to another boss, also a Chinese, and so I was brought out by ship to Singapore. I knew this third boss only by the name of Ming. I knew no more than that. He was very satisfied and pleased with me because of the profit my body and the service I gave earned for him.

My fourth boss was a Singapore Japanese. He had a great passion to own me. The bargaining went on for quite a while. Finally I was bought for seventy-five Singapore dollars, the highest price ever paid for a Japanese public-woman in Singapore. I was indeed proud that my body was more highly valued than those of the Sundanese public-women, who occupied the highest position and were the most expensive in Southeast Asia's pleasure-world.

But my pride in this didn't last long—only five months. My boss, the Japanese one, came to hate me greatly. He beat me often. He even tortured me once with burning cigarettes because my customers were declining in numbers. Such was indeed the fate that could befall even the most famous prostitute: syphilis. And what I had caught was no ordinary syphilis. In this cursed world of prostitution the disease was called Burmese syphilis. I don't know why it is called that.

It was famous for being incurable, and the men were ruined and destroyed more quickly and more painfully. Women could go without feeling anything for a long time.

So I was sold for twenty-five dollars to a Chinese boss, my fifth boss. He took me to Betawi. Before the sale took place, my old boss took me into a room. He beat me on my chest and back until I fainted. After I regained consciousness I was stripped naked and parts of me underwent acupuncture so as to kill all physical sexual desire. His name was Nakagawa. I was handed over to my new boss the next day.

On the first day with my new boss, he wanted to try me. I refused. If he found out I had that accursed disease, I'd suffer still more torture. Perhaps I'd be killed. It was nothing unusual for a prostitute to be killed by her boss and the corpse destroyed or hidden who knows where. A prostitute without a protector is a weak being. Especially too as I knew that symptoms of my declining sexual desire were beginning to show. I asked my boss to hire an acupuncturist. Three times I received treatment and my sexual desire began to revive. Yet I still refused to be tried out by my boss. He was lucky he didn't try to force me.

It was only three months later that my boss found out I had a disease. He was angry. I could tell only from his face and the tone of his voice because I understood no Chinese. My customers dwindled away. People avoided my body and he became annoyed and bad-tempered. Night and day I prayed that he wouldn't torment me. No. He could torment and torture me as long as he didn't steal my savings.

170

I hoped to return to Japan next year and marry Nakatani, who was waiting for me to bring home some capital.

My boss didn't torture me, and didn't steal my savings either. When I changed hands to become the property of Babah Ah Tjong at a price equivalent to ten Singapore dollars, he gave me half a guilder and said in broken Japanese:

"Actually I wanted to make you my concubine."

It was such a disappointment to hear those words. A concubine's life was not so harsh as a prostitute's; you could live reasonably, and were freer than the wife of a Japanese youth who hoped for capital from his future woman. What could be done? This accursed disease had taken root with me.

Babah Ah Tjong lusted greatly after me. I tried to repulse him, afraid of some new disaster. If I was exposed again, the value of my body might have fallen as low as five dollars, and I'd have become street rubbish in someone else's country. So I asked him to hire another acupuncturist. This one guaranteed I could be cured over a month with ten punctures each evening. Babah objected to the time it was going to take, and to the expense. He allowed only one treatment, as an experiment.

Before leaving for Surabaya, I ran out of excuses to refuse my boss. I was used by him alone until I was put in his pleasure-house at Wonokromo, and I had the best room.

If he was there, he almost always stayed in my room, not in any of the other fourteen rooms.

Babah didn't seem to be infected by the disease. I no longer had to worry; I was happy. There were men who were immune to the diseases of the pleasure world. Or perhaps the acupuncture treatment had made me less contagious. Who knows? And my price might go up again. Yes, I thought, who knows? If Babah had taken me as a concubine I would have been grateful and would have served him as well as any concubine could. If not, eleven months would have been enough, and I could have gone home. At the very least I would have been able to afford to redeem myself from my last boss.

That month passed. Babah became infected with Burmese syphilis too. He didn't know; he wasn't acquainted with that strange disease. He didn't accuse me straight away because there he had many other women. Neither of us could speak to each other either.

I knew he'd been affected by the disease when, one day, he lined up stark naked all fourteen of his prostitutes, all nationalities, and questioned them one by one about their diseases. In his right hand he carried a leather whip and with his left hand he measured for suspicious temperatures in those poor women's vaginas.

As a Japanese, I was the only woman he did not suspect. In the pleasure-world of this earth, Japanese prostitutes are considered the cleanest and the cleverest in looking after their health. Everyone assumes they are free of disease. So he didn't check on me.

Three of them were taken out from the line. Ah Tjong ordered the other women, except me, to tie those three up. They were gagged. Ah Tjong himself beat them with his leather whip, and there was no noise from their gagged mouths. They were my victims. And I remained silent.

It's hard being a prostitute. If you contract a disease you must quickly report it to your boss, who will then oppress you. It's best to stay silent until he finds out for himself. But the torment and oppression will still come.

After the three women recovered from their maltreatment they were sold to a Singapore middleman to be taken to Medan, in Sumatra. I still remained unaccused in Ah Tjong's pleasure-house. While I was servicing only him, I wasn't too tired. I was getting back my health and energy. Also my beauty.

Almost every wealthy Chinese has his own brothel, his own pleasure-house. In Hong Kong, Singapore, Betawi, or Surabaya, they all have the same custom, namely, to take turns visiting each other's places. So one day it was Babah Ah Tjong's turn to receive everyone.

Babah's clapping called me early in the morning. I went out. There was supposed to be gambling that morning. The afternoon and evening would be for taking pleasure. Some guests had already arrived and were playing cards, mah-jongg, or billiards in the front room.

I was already uneasy. It mustn't happen that Babah Ah Tjong surrender me to one of his guests. Who knew if they might like Japanese women very much? How many must I service if Babah allowed it?

It turned out that he did indeed order me to receive a guest: a tall, big-bodied youth, strong and handsome, healthy and attractive—of European descent. His name: Robert. My heart was moved and saddened to think of his future. I saw in a glance that he was a green one without much experience. Who wouldn't have felt compassion knowing a boy as young as that must contract such a disastrous disease if, in a little while, he desired my body. He'd carry it all his life, perhaps he'd be deformed by it or even die an early death?

I studied the look on Ah Tjong's face. Was he joking or serious? It seemed he had no worries about giving me to Robert. Then I understood: He knew that it was I who had infected him with the disease. Soon he would sell me to someone else, or he'd force me to redeem myself by paying who knows how many tens of dollars. I felt very, very sad that morning.

After Babah took Robert and me into the room and locked it from the outside, I knew I had to work, and to work as well as possible. I had to throw off all my sadness and anxiety.

Robert sat on the divan. I quickly knelt before him and took off his boots. So early in the morning! His socks were dirty and looked as if they had not been cared for properly. I took sandals from the wardrobe. None were the right size. His feet were very big. What could be done? It was only then I pulled his socks from those strong and firm legs of his. I just placed the sandals in front of him. I didn't put them on him. Such sandals, made from rice stalks, would be crushed under his feet.

He didn't put them on himself either. He seemed to be a person who thought a lot.

Robert said nothing, but only stared at me, and everything I did, with amazement.

I took off his shirt, which had two pockets. He was silent. I knew both pockets were empty. I invited him to stand up and I took off his riding trousers. I folded them and hung them up in the wardrobe, though I did so unwillingly as they were filthy and stank. His underclothes looked as if they hadn't been changed for a week. So dirty. He seemed rather embarrassed.

That was the youth Robert. He had nothing besides his youth and his health, his handsomeness and his own lust. I began to think again: Why was Babah giving me to this boy who had nothing? Perhaps he wasn't going to sell me, or force me to redeem myself because of the accursed disease. Maybe he didn't know about my disease after all. These thoughts made me somewhat happier and calm again.

From inside the wardrobe I took out one of the Majoor's kimonos. I took off Robert's underclothes and put the kimono on him. He sat silently. I gave him a goblet of special wine to strengthen him. I didn't want him to have any regrets in the future when he contracted the disease that would nest in his body forever. Let him have beautiful memories from that which would one day cause him unlimited misery, a beauty and enjoyment that was his right.

He swallowed the wine, all the time watching me in amazement. I kept talking gently so as not to disturb his mood. That was one of the many aspects of my work as a prostitute.

Of course he didn't understand a single word. Even so I did not say a single nasty word. And what man didn't like to hear a Japanese woman talk and pronounce her words? And watch her way of moving and walking? And enjoy her services inside and outside the bedroom?

At eight-thirty in the morning we got into bed. Robert refused lunch. He was very strong. His body was bathed in sweat, which made it feel as if he was made from molded copper. He didn't once let go of me. His movements were awkward, the movements of an inexperienced youth. If it wasn't for the special strengthening wine, he would have bled and not been able to become erect by himself. Let it be. Not long and his tremendous body would be destroyed and broken. Everything he had would be destroyed: his youth, handsomeness, strength. Ah, ah, ah . . . blessings that not everybody receives. So I pricked him on parts of his body just as the acupuncturist had done to me. He didn't know what I was up to, yet he bore it like a retarded little child, and I did it all while in his powerful embrace.

At four o'clock in the afternoon he let me go and got down from the bed. I too got down and wiped his sweat-covered body several times with wet towels and rosewater. Five towels! All his energy was gone. His strength and handsomeness disappeared, like old clothes piled on a chair. He asked for his clothes. I fetched them and piece by piece put them on him, including his filthy, smelly socks and his heavy leather boots. After that I scrubbed his hair and I massaged his head so it would be free from aches. I combed his hair neatly. Only then did I myself dress after scrubbing my body with a wet towel too.

He seemed very satisfied. He still availed himself of the chance

174

to grab my arm and sit me on his lap and he spoke slowly and with a deep voice. I didn't understand what it meant but I liked listening to the deepness of his voice. And I struggled to free myself, afraid that his lust would be awakened again. I hadn't breakfasted or eaten lunch. I too would have been damaged if I'd serviced him again. Perhaps his stomach was empty too.

He was so pale, as if he'd just gotten over an illness. I couldn't bear to look at him. I fetched some more special wine for him so there'd be some blood in his face. Then I took him outside the room.

But he hesitated and stopped in the middle of the doorway. Suddenly he turned and came inside again, embraced me and kissed me lustfully. Respectfully and politely I pushed him outside and locked the door from inside. I was very tired.

Below is the evidence of Babah Ah Tjong at court, spoken in Malay, and translated into Dutch by a sworn translator. I put it into order as follows:

At the time, I was in the office of my pleasure-house. About four in the afternoon the bell from the emperor room rang, signaling a request to unlock the door from outside.

I myself went to look after it. Sinyo Robert was indeed my very special guest. It's very strange that you ask why. He was the son of my neighbor, and it had long been our custom to be friendly with our neighbors. Especially as one day Sinyo Robert would be my full neighbor, not just the son of my neighbor.

He came out. His face was pale. Everything attractive about him had dissolved away. He almost couldn't lift up his head. He didn't seem to know his limits. He was somebody who one day would surrender all his body and soul to lust. Even so, he looked satisfied. That was evident from his lips, which bloomed with an unforced smile. Naturally I was happy to see this.

"Nyo," I addressed him. "We will be good neighbors beginning today and forever, yes?"

He suddenly looked at me with eyes wide open and full of suspicion. He shriveled up, frightened. Someone experienced like me could tell immediately: He realized he'd have to pay out a lot of money for the pleasure he had just taken.

"Let me sign the bill," he said hesitantly.

"Ai, Nyo, we'll be good neighbors. Sinyo doesn't need to pay anything. Don't worry. Who knows, one day we could become partners? In short you're welcome to return any time you like. You can use any room you like, as long as it's unlocked, without any time limits, night or day. You can choose anyone you like. If the front doors and windows are locked Sinyo can come in through the back door. I'll tell the gardener and the watchman."

His hesitation disappeared. Then he answered:

"Thank you very much, Babah. I never guessed Babah would be as good to me as this."

"Sinyo should have come here a long time ago. Only now you come."

"I'll come back, of course."

"Of course!"

I was his neighbor, I could never have turned him away. Especially as he was a youth still in his prime. So I not only had to think of ways to give him the chance to vent his lust but in the end I had to surrender Maiko to him until he was fully satisfied and bored.

He took leave to return home.

"It's already evening," he said.

And I didn't obstruct him in any way. I took him to my office first. His eyes went wild again when he saw the other women. He had changed; he was no longer the embarrassed youth of the morning. I pretended not to see—if I gave him another woman then I'd be breaking all the rules. So I called a woman, a barber, and ordered her to cut his hair according to my instructions.

Sinyo didn't refuse. She cut his hair in the Spanish style, with the part in the middle. His hair was rubbed with special hair oil, the most expensive. Then I ordered him to drink some palm wine from my private stock.

"So now Sinyo looks fresh again?" I said.

But that wasn't all. I gave him a dollar. A pure white dollar, like the sun, without fault. He accepted it shyly, nodded in thanks, silently.

"Babah is indeed the very best of neighbors."

I escorted him out through all the guests. Several stopped to ask for Maiko. Sinyo frowned and I refused them all. I accompanied him out to the yard. Only after his horse had reached the main road and turned left did I go inside again to find Maiko.

After that I don't know what happened to Sinyo.

And below is the story I've put together from the words of Nyai and Annelies about Robert Mellema:

At two o'clock Annelies awoke from her sleep. Her temperature had subsided. She asked whether Robert had returned.

"Not yet, Ann. I don't know where he could have got to."

Nyai was already very annoyed and angry with her son. She ordered Darsam not to leave his post. The delivery of the milk, cheese, and butter to the town was handed over to other drivers. Even the supervision of work out in back was delegated to people not really ready to become foremen.

"Let me wait for him out front, Mama," urged Annelies.

"No. Waiting inside and outside is the same. It's better in the front room with Mama."

Nyai supported Annelies and they sat together on a chair.

And still Robert didn't come. The sound of the pendulum clock disturbed the expectant atmosphere. Every now and then Nyai checked the yard. Her eldest still hadn't appeared.

"How could this happen, Ann, you falling so madly for him after just a few days? He should be the one falling madly in love with you."

Annelies didn't answer. Mama's words seemed to hurt her.

"I'll fetch some food, yes?"

"No need, Mama." But Nyai went anyway and fetched two plates of rice, meat, and vegetables, spoon and fork and drinks.

Nyai ate, spoon-feeding Annelies at the same time.

"If you don't want to chew, then just swallow," she ordered.

And Annelies didn't chew anything, just swallowed. Still no sign of Robert.

Twice Nyai called Darsam to serve customers. And Annelies sat silently gazing into the far-off distance—very far off.

Two hours passed.

"Ah, the crazy child is back at last!" exclaimed Nyai.

Only then did Annelies focus her vision on the main road.

"Darsam!" Nyai called out from her place. When Darsam

came, she continued: "Lock the office door. You stand here." She pointed to the door that joined the office with the front room.

Robert rode in on his horse, calmly, unhurriedly. He stopped in front of the steps, let his horse go free without tying it up, walked up the steps, and stood before Nyai and Annelies.

Nyai raised her eyebrows when she saw her son's haircut with the new part in the middle. She noted that he wasn't perspiring. There was no dust on him either. The horsewhip was gone. He wasn't wearing his hat. Who knows where they were.

"That part in the hair," whispered Nyai. "That paleness. . . ." She covered her mouth with her hand. "See Ann, see what sort of brother you have. Your Papa was like this when he came home from his wanderings, just like this. That smell of perfume . . . the same. If he opens his mouth, perhaps there will be that same smell of palm-wine as five years ago."

And Nyai didn't say a word to Robert.

Annelies looked at her brother with unfocused eyes. Darsam stood silently. Seeing that no one else was beginning to speak, the fighter from Madura cleared his throat. And as if he'd received an order, Robert looked up at Darsam, then shifted his look to his mother.

"The police don't know where Minke was taken. They don't recognize that name at all."

Nyai stood up, enraged. Her face was scarlet. She pointed her finger at her eldest son and hissed:

"Liar!"

"I've been everywhere seeking an explanation."

"All right, enough. You don't need to say anything. The smell on your breath, your perfume, that hair-part . . . the same as your father five years ago and ever since. Look well, Ann, that was the beginning of your father losing all his sense of direction. Go, get out, you liar! I have no son, a cheat and liar."

Darsam, standing before the office door, coughed again.

"Never forget this day, Ann. This was how your father arrived that time, and from then on I had to think of him as gone from my life. So today it is the same with your brother. He is following in the footsteps of Master." Annelies did not respond. "So be it. Because of this, Ann, you must be strong. You must be strong, otherwise you will become a plaything, and will be played

178

with by people like him. Stop crying. Do you want to follow your father and brother?"

"Mama, I want to be with you, Mama."

"So don't indulge yourself. Strengthen your heart."

Annelies was struck silent on seeing Nyai suffer her greatest disappointment.

The horse in front of the house neighed. Robert came out of his room in clean clothes, neat and dashing. He walked out of the house quickly, paying no attention to his mother, sister, and Darsam.

He made no attempt to tie up the horse either. He just walked away.

Since that day Nyai's eldest son hardly ever set foot in his family's house.

11

I woke up at nine o'clock in the morning with my head aching. There was a throbbing and pounding above my eyes. It was as if a palakia tree seed had, unbeknownst to me, penetrated my skin, and was now growing roots in my brain in order to turn itself into a tree inside my head.

And I remembered the newspaper reports hailing the most powerful medicine discovered in the history of humanity, a medicine that would do away with headaches forever. They said the Germans had discovered it, and it was called *aspirin*. But so far it was no more than a report. It was not yet to be found in the Indies, or at least I didn't know of it. Indies, a country that can do nothing but wait upon the products of Europe!

Mrs. Telinga had compressed my head several times with brown-onion vinegar. The whole room smelled of vinegar.

"Perhaps there's a letter for me, ma'am?"

"Ha, only now Young Master asks about letters. Before, you were never interested in reading them. You've changed. Perhaps, though, there is one. The person was here a while ago. I said you were sleeping. I don't know his name. Perhaps he has gone. I said

180

to him: Young Master Minke is now staying at Wonokromo, isn't he? He didn't seem to take any notice, but rather took leave to visit next door for a moment, to Marais's house."

The boarding house was still. The other boarders had all left for school.

And that good-hearted woman pulled the table over close to my bed, then put some hot chocolate and fried coconut patties on it.

"What would Young Master like to eat today?"

"Does ma'am have shopping money?"

"If I run out, I can always ask Young Master for more, yes?"

"Perhaps a police officer has recently been here asking about me?"

"There was somebody. Not a police officer. A young man of Young Master's age. I thought he was a friend of yours, so I told him what had happened."

"Indo, Pure European, or Native?"

"Native."

I didn't question her further. I reckoned he was none other than that same police officer.

"So what will Young Master eat tonight?"

"Macaroni soup, ma'am."

"Good. This is the first time you've wanted macaroni soup. Do you know how much one packet costs? Five cents, Young Master. So . . ."

"Two packets will be enough, no doubt."

She laughed, relaxing when she received the shopping money of fifteen cents, then rushed hurriedly back to her kingdom: the kitchen.

That morning everything was still. Occasionally the bell of a passing buggy could be heard. It was only within my mind that there was great activity: Murderers and future murderers formed long lines with all sorts of faces, all sorts of figures. Even Magda Peters appeared in my mind with a threatening, unsheathed dagger. Magda Peters—my favorite teacher! It was as if I had gone mad, just because of what someone had said. Come on, I shouldn't be so afraid of something whose truth and real context are so uncertain! I, an educated person? Even if the report was true, is it proper that I surrender to this accursed fear?

You will have suffered twice, Minke, if this report turns out

181

to be true. First of all, you're already afraid. Secondly, you'll be killed. It's enough to suffer once, Minke. Once is enough. Get up then. Why must you suffer twice? You're too stupid to be called educated.

These thoughts made me laugh at myself. So I got up from the bed, stood tottering for a second, and started walking to the back of the house. Everything seemed to be moving. I grabbed for the back of a chair. I got my vision under control again and left the room. I didn't end up going out to the back but sat in the front room and began to try to read the newspapers.

My headache subsided somewhat, but the smell of the onion vinegar continued to be a real annoyance.

A pampered body, I said to myself.

Finally I did go out to the back and had a hot bath despite the protests of the garrulous Mrs. Telinga. How fond she was of me, that childless lady. She was an Indo who was more Native than European, without a single remaining trace of beauty, fat like a pillow. Even though her Dutch was terrible, it was still her day-to-day language; it was the language used by her family. She had never set foot in a school: illiterate. Her adopted child was a male mongrel dog. It was very clever at stealing fish from the market, which it would do two or three times a day. It brought the fish home to its adopted mother and she would grill it for the dog. After eating, the dog would go to sleep in the middle of the door, only to wake up and set off to steal again. This adopted child never barked at strangers, but would look them over with slowly blinking eyes as if waiting to be barked at first.

After getting dressed and combing my hair, I went to Jean Marais's house. The picture of May's mother fighting the soldier wasn't finished. He was putting an all-out effort into it. He wanted it to be his best work.

May sat on his lap enjoying being spoiled. She had missed me over the last few days. I usually brought her some sweets. This time there wasn't anything in my pockets.

"We're not going for a walk, Uncle?"

"I'm not feeling well, May."

"You're pale, Minke," Jean Marais admonished me.

"I didn't notice," May said, using French too; then she got up from her father's lap and looked at me. "Yes, Uncle, you're pale."

"Not enough sleep," I answered.

"Since you moved to Wonokromo all sorts of things have been happening to you, Minke," Jean reprimanded. "And you haven't been out looking for any new orders since then either."

"If you knew what I've been through recently, Jean, you wouldn't be able to bring yourself to talk like that. Truly."

"You're in trouble again," he accused. "Your eyes aren't calm as they usually are."

"Come on, can you really know what's happened to somebody from their eyes?"

"May, buy some cigarettes for me, please."

And the little girl went outside.

"Now, Minke, tell me what troubles you now."

Naturally I told him of my suspicions about Fatso. That I felt there was somebody waiting for the opportunity to kill me (and there is only one of me!). That I felt that everywhere there were people spying on me, ready to swing their machetes into action against my body.

"Just as I expected. This, of course, is the risk you face when you go to live in the house of a nyai. You once joined in condemning the nyais and their morality. And what did I say? Don't sit in judgment over something about whose truth you are uncertain. I suggested you visit there two or three times, to observe things as an educated person."

"I remember, Jean."

"Well, you have been there. But you didn't just visit, you stayed."

"Yes."

"You lived there not just to investigate how the common view of nyais matches with reality; you decided to act out that common view yourself—dragged down to a low and shameful level of moral behavior. Then you receive threats, who knows from whom. No doubt from that quarter with the most at stake: and whom you have now challenged. Now you fear someone is out to get you, Minke. But it is your own guilt that is pursuing you."

"What else Jean."

"Am I wrong?"

"It's very possible you are right."

"Why only possible?"

"That is, if it were true I had acted in the shameful manner you imply."

183

"So you haven't?"

"No, not at all."

"At the very least I'm happy to hear that, Minke, my friend."

"And it's also clear that Nyai is no ordinary woman. She is educated, Jean. I think she's the first educated Native woman I have met in my life. Wonderful, Jean. Some day I'll take you there to meet her. We'll take May. She'd like it very much there. Truly."

"So why does someone want to kill you if you have done nothing wrong? You're educated, try to be true to your conscience. You are among the first of the educated Natives. Much is demanded of you. And if you can't deliver, the educated Natives after you will grow up more rotten than you yourself."

"Quiet, Jean. Don't talk such nonsense. I'm in trouble."

"It's only your imagination."

May returned with a packet of corn-leaf cigarettes and Jean quickly started smoking.

"You smoke too much, Jean."

He only laughed. On that day my French friend did not make me happy at all. He was wrong. And I was being reproached with unfounded allegations. Father too had made similar accusations at the beginning of our meeting. Now Jean Marais seemed not to believe the truth of what I was saying: He too, in the end, suffered the old prejudices. He thought I'd been defeated, dragged down to shameful behavior. It felt as if there was no use in continuing the conversation.

I took May's hand and we went home together. We sat on the long bench on the front porch.

"Why don't you go to school, May?"

"Papa likes me to wait on him while he paints."

"So what else do you do?"

"Watch Papa paint, just watch."

"He doesn't talk to you?"

"Of course he does. He said that under the clump of bamboo it should be cool because the wind whistles through all the time. But that poor person being stamped on by the soldier, Uncle!?"

She didn't know that the woman being trodden upon was her own mother.

"Come on, May, sing!" The child straight away started singing her favorite song. "A French one, May. I already know all the Dutch songs."

"French?" She tried to remember. Then: "Ran, ran pata plan! Ran, plan, plan," from "Joli Tambour." "You're not listening, Uncle, come on!"

My eyes were observing a fat man wearing a sarong and sitting under a tamarind tree on the other side of the road, next to a chilled-fruit seller. He wore a peci, but wasn't wearing sandals, let alone shoes. His shirt was made of calico, and he wore loose, black trousers and a wide, leather belt. His shirt was unbuttoned. His figure and skin and his narrow, slanted eyes meant he could not fool me. Perhaps he was to be my future murderer. Fatso! Robert's man, now that he had failed in enlisting Darsam.

Every now and then, while eating his fruit, he glanced over at the two of us.

"Call Papa, May."

The little girl ran off. And Jean appeared with his tall, thin body, walking lamely on his armpit crutch. He sat down beside me.

"I don't think I'm wrong, Jean, that's the one. He followed me from B——. Only his clothes are different now."

"Ssst. It's just your own imagination, Minke," he scolded me instead.

Just at that moment Mr. Telinga arrived from somewhere. In one hand he was carrying a basket—who knows what its contents were—and in the other hand he held a length of steel pipe, obtained from who knows where.

"Jean, Minke, what's going on? You two sitting out here together so early in the morning?" He greeted us in Malay.

"It's like this." Jean Marais began the story about my fears. Then he pointed with his chin to the man I reckoned was Fatso.

The new arrival put his basket down on the ground: It proved to be full of young kedondong fruit. The steel pipe remained in his hand. His wild eyes were aimed across the road.

"Let me have a close-up look. Come on, Minke, you're the one who knows him. Perhaps it is him. Let me bash in his head if need be." I walked along behind him and Jean Marais followed us, limping.

It became clearer as we approached that it was Fatso. He was no doubt spying on me that very moment. And he was pretending not to notice us coming closer. He went on enjoying his fruit, though I could see his eyes glance about vigilantly. That he was disguised only strengthened my suspicions.

"It's him all right," I said without hesitation.

Telinga approached him threateningly, the steel pipe still in his hand. I myself no longer knew what I wanted to do. Jean Marais was still limping along behind us.

"Hey, man," Telinga snapped in Javanese, "are you spying on my house?"

Fatso pretended not to hear and continued eating.

"So you're pretending not to hear, eh?" snapped the pensioned Indies army soldier, this time in Malay. He grabbed the plate of fruit and threw it on the ground.

It appeared that Fatso was not scared of Indos. He stood up, wiped his chili-paste-covered hand on a piece of tamarind trunk bark, swallowed down the remains of his fruit, bent over and washed his hands in the fruit-salad seller's bucket of water, and only then spoke, calmly, in High Javanese:

"I am not spying on anything or anyone." He tried to glance across at me and smiled.

What did he think he was doing? The impudence! He smiled at me. My future murderer! He smiled!

"Get out of here," shouted Telinga.

The fruit seller, an old woman, frightened, moved away. Some distance away people began to congregate, no doubt curious to know why a Native dared confront an Indo European.

"I eat here almost every day, Ndoro Tuan."

"I've never seen you. Go! If not . . . " He swung the steel pipe about.

And Fatso was still not frightened. He didn't lift up his head and just bowed, his eyes wary.

"There has never been any ban on eating here, Ndoro Tuan," he retorted.

"You dare argue back? Don't you know I'm Dutch and was in the Indies army?"

Fatso was no doubt a fighter. He wasn't afraid of an Indies army Dutchman. Perhaps he was a Malay-style fighter, or skilled in Chinese martial arts.

"Even so, there is no police ban. There has not been any public announcement of any ban, Ndoro Tuan. Allow me to just sit here and eat my fruit. I haven't paid yet, either," and he was about to sit down again.

I became suspicious when I heard him talk about bans. He

186

knew about regulations. Telinga must be more careful. But the former Indies army soldier, who only knew of violence, had already swung his arm, striking towards the side of Fatso's head. Fatso parried the blow, but did not return the attack.

"Enough, enough." Jean Marais tried to intervene.

"Don't keep on, Ndoro Tuan," appealed Fatso.

Telinga lost his temper: This person dared defy him! He no longer cared who Fatso was. His pride had been wounded. His right hand swung a death blow down towards Fatso's head. And the rebel calmly moved out of its way. Telinga was thrown forward by his own blow as it missed its target. In fact Fatso could have easily punched Telinga in the ribs, but he didn't. Each time Fatso successfully avoided Telinga's blows, Telinga become angrier, and he attacked again and again. Fatso retreated, then retreated a little more, and then ran off. Telinga chased after him. Fatso disappeared down a narrow alley, where there were heaps of rubbish piled up.

"Telinga's crazy!" Jean Marais frowned. "He thinks he's still in the army."

And the person Jean was frowning at went off hunting after Fatso, and disappeared down the same alley.

"What's all this in aid of? Come on, let's go home, Minke. You're the cause of all this."

He wouldn't let me support him. May and Mrs. Telinga rushed about asking us what was happening. No one would tell them. The two of us sat and waited for the hothead to return. Anxiously, of course.

Ten minutes later, Mr. Telinga returned, bathed in sweat, red-faced and short of breath. He threw himself down into a canvas deck chair.

"Jan," his wife complained, "what's wrong with you? Have you forgotten that you're an invalid? Looking for enemies? Do you think you're still young?" She approached her husband, seized the pipe from his hand, and took it inside.

Mr. Telinga didn't speak. It was as if there was a secret agreement between us. And no one regretted all that had happened more than myself. Deep in my heart I thanked God that no catastrophe had happened. And I felt it was fortunate that I hadn't told them Darsam's story. I would have really been blamed for everything then.

187

"Young Master is still ill," called out Mrs. Telinga from inside. "Don't sit outside in the breeze. It's better you get some sleep. Food will be ready in a little while."

"Go home, May," ordered Jean, and May went home.

The three of us sat silently until Telinga had got his breath back.

"Let's forget everything that has happened," I proposed. What if the police became involved. I'd truly be the shameful cause of it all. "My head is aching again, Jean. Excuse me Mr. Telinga, Jean . . ."

Once in my room I became even more convinced: It had been Fatso and he had been spying on me. He was obviously one of Robert's men. Darsam's story could not be dismissed as nonsense. Be careful, Minke!

For the first time ever, I locked my door and windows while it was daylight. I got ready a hard wooden stick, formerly a mop handle, and placed it in a corner. I could grab it at any moment. Even if only at a basic level, goat-class, we say in Malay, I too had studied self-defense when I was in T——.

Being an educated person, there was something I had to accept: Someone wanted to take my life, but I could not report it to the police. It would be unwise to involve Nyai and Annelies, or Father, who had just been made a bupati, or, most of all, Mother. I had to face all this silently, but with vigilance.

Four days passed and my headache still did not disappear. I certainly wasn't getting enough sleep. And every day milk arrived. But there was no news from Darsam.

I had been away from school too long. The doctor had given me a certificate to stay away three weeks. The palakia seed had grown into a full-grown tree inside my head, without permission from its one and only legitimate owner, me. You're right, Palakia Tree in My Head, I have to forget Nyai and Annelies. I must cut all relations with them! There is no point in maintaining relations. Only trouble will come of them. My life would suffer no great loss. I would not catch leprosy by not knowing this strange and forbidding family. I must become well again. Go after orders as before. Write for the papers. Graduate from school as was hoped for by so many people. Whatever else, I still liked school. To mix openly with all my friends. To be free. Obtaining new, unlimited knowledge and learning. Absorbing everything from this earth of mankind from the past, the present, and the future. At the end of next month, Magda Peters was going to hold discussions, lighting up this earth of mankind from every possible angle. And I was ill like this.

I had wasted these emergency holidays: a clot of time packed with tension. Often I thought: Should someone as young as I have tension such as this? As often I replied: No, not yet. Miss Magda once told us the story of Multatuli and his friend, the poet-journalist Roorda van Eysinga: They lived in tension because of their beliefs and their struggles to improve the fate of the people of the Indies, against both all-European and all-Native oppression. For the sake of the peoples of the Indies, who knew nothing of the world, those two ended up in exile, without comrades who visited them, without a single hand stretched out in aid, Minke. Read Roorda van Eysinga's poem, written under the pseudonym Sentot, "The Last Days of the Hollanders in Java." Every word was filled with the tension of somebody calling out in warning.

Multatuli and van Eysinga came under great pressure because of profound actions. And the pressure that I was under today? Nothing more than the result of the stumblings of a philogynist. I must let go of Annelies. I must be able to do it! But I still couldn't convince myself. A girl as beautiful as that! And Nyai—such an inspiring and impressive character—a regal woman of great powers of betwitchment. Yes, yes: "If attracted, no limits to one's praise; if hated, no limits to one's finding fault."

Slowly I began to understand: All this was the result of my reluctance to pay the price of entering the world of pleasure, the world where dreams become reality. Multatuli and van Eysinga paid the price but they wanted nothing for themselves. What do my writings mean compared to theirs? Everything I hope and lust after is for myself. The thought shamed me.

Yes, I must let go of Annelies. Adieu, ma belle! Happy separation, oh dream, which I'll never meet again, not at any time nor in any place. There are things more important than a girl's beauty and a Nyai's charisma. There is no point to a meaningless death. And my life and my body are my basic capital and the only life and body I have.

This decision chased away the headache, though not all at once. That's how all illness develops: It comes on suddenly, but disappears lazily. The palakia tree stopped spreading its roots and seedlings. Then they too died, but only because of the arrival of a letter from Miriam de la Croix. Her writing was fine and small, neat.

She wrote:

My friend, You have no doubt arrived in Surabaya safely by now. I've waited for news from you but none has arrived. So it is I who give in.

Don't be surprised, but Papa is very interested in you. He's already asked twice whether there have been any letters. Papa wants very much to know how you're progressing. He was impressed by your attitudes. You, he said, were a different type of Javanese, made from different material, a pioneer and innovator at one and the same time.

I write this letter gladly. Indeed, I feel honored that I am able to pass on Papa's opinions. Mir, Sarah, he said to us another time, that will be the face of Java in the future, a Java which has absorbed itself into our civilization, no longer shriveling up like a worm struck by the sun. Excuse him, Minke, if Papa uses such a coarse metaphor. He does not intend to insult. You're not angry, are you? Don't be angry, my friend. Neither Papa nor the two of us have any evil thoughts towards Natives and especially not towards yourself.

Papa feels moved when he sees how the Javanese people have fallen so low. Listen to what Papa has also said, even though he still uses that earlier coarse metaphor: Do you know what is needed by this nation of worms? A leader who can give them dignity once again. Do you follow me, my friend? Don't get angry before you know what I mean.

Not all Europeans have been participants in, and causes of, the fall of your people. Papa, for example, even though he is an assistant resident, is not one of them. Indeed, he is unable to do anything, just as Sarah and I are also unable to do anything, though we all desire so greatly to stretch out our hands to help. We can only guess now at what we must do. You yourself are fond of Multatuli, aren't you? That writer, so glorified by the radicals, has indeed been of great service to your people. Yes, Multatuli, as well as Domine Baron von Hoëvell and another person also, whom your teacher perhaps forgot to tell you about, that is, Roorda van Eysinga. But all these people never spoke to the Javanese, only to their own people, the Dutch. They asked Europe to treat your people properly.

But now, says Papa, at the close of the nineteenth century, nothing that they did is of use today. According to Papa, it is the Natives themselves who must now act. So when we talked about Dr. Snouck Hurgronje that day, it was no coincidence. That particular scholar holds an honored place in the thinking of our family. We praise Association Theory, at which you laughed that day. So please understand, my friend, why Papa is so interested in you. Indeed, Papa, and the two of us too, had never met a Javanese like you. You, he said, were totally European. You showed no sign of the slave mentality that the Javanese developed during the era of their defeat, from the time the Europeans set foot on this, the land of your birth.

On those still nights in this big and empty building, if Papa is not tired, we like so much to sit and listen to his explanations about the fate of your people. How they gave birth to hundreds and thousands of leaders and heroes in their struggle against European oppression. One by one they fell, defeated, killed, surrendering, gone mad, dying in humiliation, forgotten in exile. Not one was ever victorious in war. We listened and were moved, and became angry also to hear how your rulers sold concessions to the Company, benefiting no one but themselves. It was a sign that their character and souls were being corroded. Your heroes, according to Papa's stories, always emerged out of a background of selling concessions to the Company; and so it was over and over again, for centuries, and no one understood that it was all a repetition of what had gone before, and that as time went on the rebellions became smaller and more and more stunted. And such is the fate of a people who have thrown all their body and soul and all their material

191

wealth into saving a single abstract concept called honor.

According to Papa, the fate of humanity now and in the future is dependent on its mastery over science and learning. All humanity, both as individuals and as peoples, will come tumbling down without such mastery. To oppose those who have mastered science and learning is to surrender oneself to humiliation and death.

So Papa agrees with Association Theory, that is the one and only road for Natives. He hopes, and so too do both of us, that one day in the future you will sit together, as an equal with Europeans, in advancing this people and this country. You have already begun this yourself. You understand what I mean. We love our father very much. He is not simply a father, but also a teacher who leads us in our efforts to see and understand the world. He is a friend, a mature man with great authority, an administrator who hopes for no profit from the woes and tribulations of those under him.

Let me tell you what he said after you left us following that first visit. You left feeling angry and irritated, didn't you? We could understand because we knew that you didn't understand our intentions. Papa left us alone on purpose, so you could speak freely with us. But a pity, you were so formal and tense. As soon as you left, Papa asked us our opinion of you. Sarah reported: Minke became angry in the end, Dr. Snouck Hurgronje and his Association Theory were three hundred years behind the times, repeated just as you had said. Papa was surprised and had to ask me to explain further. Then Papa said: He is proud to be a Javanese, and that is good as long as he has self-respect as an individual as well as a child of his race. He mustn't become like the general run of his people; when among themselves they feel as if they are from a race which has no equal on the earth. But as soon as they are near a European, even just one, they shrivel up, lacking the courage even to lift up their eyes.

I agree with this praise of you, my friend. May things go well for you.

Then from the wayang orang performance building, gamelan music began to waft across to us. Papa had

192

ordered us to study your people's music. You have studied gamelan for a long time now, he said to us, and perhaps you can already enjoy it. Listen to how all the tones wait upon the sound of the gong. That is how it is in Javanese music, but that is not how it is in real life, because this pathetic people has still not found their gong, a leader, a thinker who can come forth with words of resolution.

My friend, I ask with all my heart that you try to understand these words, which you will obtain nowhere else except from my father, not even from the great scholar Snouck Hurgronje. That is why we are so proud to have a father like him. Papa is sure that the reason you like gamelan better than European music is because you were born and brought up under the swaying sounds of your great gamelan.

Minke, my friend, when will Life's gong sound? Will you perhaps be the one? The great gong? May we pray for you?

Listen again to the gamelan, said Papa once more. It has been that way for centuries. And the gong in the life of the Javanese has still not arrived. The gamelan sings of a people's longing for a messiah. Just longing after him, not seeking him out, not giving birth to him. The gamelan translates the life of the Javanese, a people who are unwilling to seek, to search, who just circle around, repeating, as in prayers and mantras, suppressing, killing thought, carrying people into a dispirited universe, which leads them astray, where there is no character. Those are the views of Europeans, my friend. No Javanese, not one, would think like that. Papa also says that if things remain like this for another twenty years this people will never find their messiah.

My friend, what will your so-saddening-a-people look like twenty years from now? One day in the future, we will go home to the Netherlands. I will go into politics, Minke. It's a pity though that the Netherlands still doesn't allow a woman to sit in the Lower House. I have a dream, my friend, that one day, when it is no longer as it is now and I can become an Honorable Member of the

Lower House, I will speak much about your country and your people. And when I return to Java I will, first of all, listen once again to your gamelan, a gamelan whose beautiful unity of sound has no equal on earth. If its theme is just the same, a longing without effort, it means no messiah has yet arrived or has yet been born. It also means that you have not yet emerged as a gong, or indeed no Javanese ever will, and your people will drown forever in the overflow of repeated tones and vicious circles. If change has taken place, I will seek you out, and hold out my hand in respect.

Friend, twenty years! That is such a very long time in this pounding, racing era; and it is, of course, also a long time if looked at as part of somebody's life. My friend Minke, this is the first letter from your sincere and well-wishing friend,

Miriam de la Croix.

As I folded up the letter I knew that tears had left blue smudges here and there where the ink had run. Why had a letter from a girl whom I had met only twice in my life made me cry? She was neither kith nor kin, not even of the same race. She had such hopes for me. While there I was totally confused because of my recent mistakes at Wonokromo. She wants me to be of value to my own people, not to her people. Is it possible there is now a new style of Multatuli and van Eysinga?

How should such a beautiful letter be answered? I had begun to consider myself a writer and had been praised by Mr. Maarten Nijman, Chief Editor of S.N.v/dD, but I felt too small to be able to equal Miriam's thoughts. Still I forced myself to answer. Thank you, and no more than thank you, poured out in so very many words, perhaps just like the pouring out of Javanese musical tones that separately headed for and waited upon the gong. In the letter I sated my astonishment that Multatuli and Eysinga, whom I had just been thinking about, should be mentioned in her letter. Perhaps, I wrote, because we live in the same liberal era, in the same era, and I ended my letter with:

My good friend Miriam, I am so lucky to have found a friend such as you. I do not know what is going to

happen in the next twenty years. I myself have never felt
that I would ever become a gong. Neither had I ever
dreamed of being even a drum; I have never had such
thoughts, perhaps would never have had such thoughts
had your beautiful and moving letter not arrived. Especially as it has come from someone who is not of my
people. Peace and well-being be with you, my sincere
Miriam. May you indeed one day become an Honorable
Member of the Lower House.

I rested my face on the table. I tried to absorb all of Miriam's
letter, trying to ensure I would never forget it as long as I lived.
Friendship is indeed beautiful. And my headache slipped away and
slipped away, and then disappeared altogether, who knows to
where. Miriam, you did not just send a letter. More than that: a
charm to rid me of tension. If only you knew: Suddenly I felt
brave, and the world became brighter and clearer. Become a gong
to be heard booming out everywhere.

"Young Master!" I raised my head. On seeing the person in
front of me the tree in my head returned, spreading its roots and
seedlings. But now more vigorously. Him. Darsam.

"Excuse me, Young Master. I must have surprised you,
you're so pale."

I tried to smile, my eyes darting to his machete and his hands.
He laughed in a friendly manner while stroking his mustache.

"Young Master suspects me," he said, "when Darsam is
Young Master's friend."

"So what's the matter?" I asked, pretending not to know
anything.

"A letter from Nyai. Noni is very ill."

I started and my eyes opened wide. He stood across from me.
I read the letter while every now and then glancing across at his
machete and his hand. Yes, it was true, Annelies was very ill and
was being looked after by Dr. Martinet. Nyai had told him of the
origins of her illness, and she requested very strongly that I come
quickly. This was also advised by the doctor. Dr. Martinet said
that without my presence Annelies had no hope of recovery; and
her illness would probably get worse and there would be compli-
cations.

"Come, Young Master, to Wonokromo now."

195

My head throbbed as if it wanted to break open. I couldn't stand up straight; my body went flabby. I quickly grabbed for the corner of the table. With my shaky vision I gazed at the fighter. Darsam grabbed me by the shoulder.

"Don't be concerned. Sinyo Robert will not be able to worry you. Darsam still stands vigilant. Come."

Miriam de la Croix vanished, evaporated into smoke, disappeared from circulation. The magical power of Wonokromo had regained control over everything. Aided by Darsam, these legs of mine carried me to the buggy waiting in front of the house.

"You're not going to take leave from the people here?"

I came to a halt. I called out to Mrs. Telinga that I was excusing myself and was leaving. She stood at the door and did not seem at all happy.

"Don't be away long, Young Master," she reminded me. "Your health."

"Young Master will recover quickly at Wonokromo," answered Darsam.

Afraid of Darsam's frightening appearance, she didn't say anything more.

"Where are your things, Young Master?"

I didn't answer. And I never found out whether I fainted in the buggy or not. All I knew was that all this had occurred as a result of Robert Suurhof's invitation, and now so many people had become involved and this young life of mine had become so tense. All the time I heard just one voice, one sentence emerging from the Madurese fighter's mouth: "This carriage and this horse are henceforth Young Master's property."

12

As soon as Darsam led me up the stairs, Nyai Ontosoroh came rushing out to greet me.

"You've gone too far, Nyo, we've been waiting and waiting so long for you. Annelies has fallen very ill longing for you!"

"Young Master was also ill, Nyai. I had to carry him here."

"No matter. If the two of them get together again, everything will be all right. The sickness will disappear."

Those words were so embarrassing, yet it already felt like they had started to work as an antitoxin, dissolving away the palakia in my head. Nyai caught me by the shoulder and whispered softly in my ear, smiling.

"Yes, your temperature is a bit high. No matter. Let's go upstairs, child. Your little sister has had to wait too long. You didn't even send news."

She spoke so gently, it went straight to my heart. It was as if she was my own mother, my beloved mother, and I no other than her little boy, under her guidance. Yet my eyes kept glancing about here and there. At any moment Robert could leap out

197

of the darkness and thrust a knife into me with those mighty muscles of his.

"Where is Robert, Mama?" I asked, as we climbed the stairs.

"Sst. You don't need ask that. He's his father's son."

Why did I become so malleable in the hands of this woman? Like a lump of clay that could be molded just as she wished? Why was there no fight in me? Even the will to fight was missing, as if she understood and had mastery over my inner self, and could lead me in the direction I myself desired?

The upstairs was far more luxurious. Almost all the floors of the corridor were covered with carpet. It felt as if I were a cat who could walk along without making a single sound. The open windows offered views that stretched far away out there into the distance. Paddies and fields and forest spread out everywhere, joined together one after the other. A small group of people were collecting the last of the harvest. The remaining paddy was still fallow, awaiting the beginning of the end of autumn.

The newspapers reported that there had been an abundant harvest that year. There was no need to import low-quality rice from Siam, even though the most fertile rice lands of east and central Java were, to all intents and purposes, producing only sugar. It was a sign, said one observer, that Queen Wilhelmina had been blessed by God as the youngest queen ever, at an age very young for a queen.

We stood in front of the bed. Nyai fixed the blanket that lay over Annelies. That girl's breasts stood out underneath the blanket. And Nyai put her daughter's hand into mine.

"Annelies, darling."

With great effort the girl opened her eyes. She didn't turn. She didn't look. Her eyes, and that effort-filled look, were swept up to the ceiling, then they closed again.

"Minke. Nyo, child, take care of my sweetheart here," whispered Nyai. "If you too are sick, then get well now. Carry my child to recovery with you." It sounded as if she was praying.

She looked at me with almost begging eyes.

"It's up to you, child. As long as my daughter recovers. . . . You're educated. You know what I mean." She bowed down as if embarrassed to look at me. Her two hands held my arm. All of a sudden she turned and left the room.

I felt for Annelies's hands under the blanket. Cold. I brought my mouth close to her ear and called her name again and again, slowly. She smiled, but her eyes remained closed. Her temperature was not too high. And I knew then: The palakia tree inside my head had been flung out, plucked out with all its roots and seeds, crashing down to who knows where.

And she was so close to me. My heart began to pound rapidly, pumping hot blood all over my body, and I began to perspire.

"Come on, haven't you been waiting for Minke to come?"

I didn't know whether it was just my imagination, or if it really happened, but I saw her nod weakly. Her eyes and mouth remained closed.

"Do you miss him, Ann? Of course you miss him. He misses you too. Truly. If you only knew how much he longs to be always near you, Ann, to make you the adornment of his life, all the world would then be his, because his happiness is you, you alone. Open your eyes, Ann, because Minke is here now, with you."

I heard Annelies sigh. Her eyes and lips remained closed.

Doesn't this girl recognize my voice anymore? So I caressed her face, her cheeks, her hair. She tilted her head and sighed once more. Was she going to die? A girl as beautiful as this? I embraced her and I kissed her on her lips. The beat of the heart in her breast seemed too slow. Her fingers moved slowly, almost stilled.

"Ann, Annelies!" I finally cried into her ear. "Wake up, Ann," and I shook her shoulders.

She opened her eyes and stared far into the distance, not seeing, and not reaching my face.

"Don't you know me anymore, Ann? Me? Minke?"

She smiled. But she still stared right through me.

"Ann, Ann, don't be like this. Aren't you happy now Minke is back? I've come. Or must I go again and leave you? Ann, Ann, my Annelies!"

She must not die here in my embrace. I stood before the bed and wiped sweat from my dripping forehead.

"Keep going, Nyo," Nyai encouraged me from the door. "Keep asking her to talk. That's exactly what Dr. Martinet advised."

I turned. Nyai was pulling the door shut from outside. Her encouragement calmed my anxiety. Annelies was not facing death. She just wasn't conscious.

I sat on the edge of the bed. Her eyes were open but she didn't see anything.

"You can't go on like this, Ann," I said, trying to convince myself as well. I pulled back her blanket. I pulled her up by her two hands. I forced her to sit up. But her body was so weak it fell back upon the pillow as soon as I let go of her. I tried again. She still couldn't sit up.

What must I do now?

Once again I kissed her upon her lips. Her hands began to move almost imperceptibly, but a little more than they had done a minute ago. I shifted her head across to my left arm. I began again to try to get her to talk.

"If you're sick like this, who'll help Mama? There is no one else. So you mustn't be ill. You must get better. So that you can work, and go walking with me. We can go riding, Ann, all around Surabaya."

I looked into her eyes, which stared off into the distance, and I could see myself in the depth of her eyes. But she still didn't see me. For a moment I thought my face was not even reflected in her eyes.

Nyai Ontosoroh came back, carrying two glasses of warm milk. One glass she put on the table. The other she brought over to me and put it up to my lips so that I would drink it down quickly.

"Drink it all up, Nyo, child, Minke." I drank until it was finished and the glass was dry. "You must be healthy and strong too. A weak and ill person is of no use to anybody." Then to Annelies: "Wake up, Ann, Minke is here with you now. Who else are you waiting for?"

Without waiting to see whether there was any reaction from Annelies, she left again.

There had been no change in the situation the next time Nyai returned, but this time she came with Dr. Martinet. I put Annelies's head down on the pillow in order to be able to greet him.

"This is Minke, Doctor, who has been looking after Annelies today," and we shook hands. Nyai observed us for a moment, then she continued, "Excuse me, I must go downstairs."

"So you are Mr. Minke, the H.B.S. student? Excellent. Happy is any young man who obtains so deep a love from so beautiful a girl," he said in mumbled Dutch.

"I have only been here an hour, Doctor. This was how Annelies was when I arrived. I'm worried, Doctor."

The forty-year-old man let go a laugh, shook his head, and shook my shoulders.

"You like this girl? Answer frankly."

"Yes, I do, Doctor."

"You have no intention of playing foul with her, eh?" He fixed his gaze squarely upon me.

"Why would I do that?"

"Why? Because H.B.S. students have always been the favorites of the girls. It has always been like that, ever since the schools were founded. In Batavia too, and Semarang also. I repeat, Mr. Minke, you have only honorable intentions?" Seeing that I remained quiet, he went on: "There is only one thing needed by this girl: you, Mr. Minke. She has everything she needs, except you."

I bowed my head. Confusion filled my breast. I had no intention of manipulating Annelies. But neither had I ever intended becoming serious about any girl. Now Annelies wanted all of me, totally, for herself. I was being tested by my own deeds. And it was my conscience that made me affirm things of which I was not yet really convinced.

"You want her to regain consciousness?"

"Of course, Doctor, I would like it very much, and would be very grateful if you make her better."

"She will regain consciousness. I have been drugging her and waiting for you to arrive. So it is your fault that she has been drugged for so long. If I had left her conscious without you here, there is no telling what would have happened. If you hadn't come back and I was forced to keep her drugged, it would have damaged her heart. It all comes back to you—you are the cause of it all."

"Forgive me."

"She has chosen you to be the one to accept all the risks."

I didn't respond. And he kept talking.

"She will be conscious again soon. About another quarter of an hour. When she starts showing signs of waking, you must begin talking to her, just pleasant, happy things. Don't speak harshly or roughly. Everything depends on you. Don't disappoint her. Don't make her lose confidence or become afraid."

"Very well, Doctor."

"You passed this year's exams?"

"Yes, Doctor."

"Congratulations. Wait on her until the drugs wear off. What is your family name, if I may ask?"

"I have none, Doctor."

He cleared his throat. His eyes swept over my face for just a moment. Then he went over to the window, and looked out onto the fields and the garden beside the house.

"Come over here," he invited without turning.

And I stood beside him at the window.

"Why do you hide your family name?"

"I don't have one."

"What is your Christian name?"

"I don't have one, Doctor."

"How is it possible that you're in the H.B.S. and have neither family nor Christian name? You don't mean to tell me that you're a Native?"

"Yes, I'm a Native, Doctor."

He glanced at me. He stood silently for quite a while, perhaps trying to convince his own heart of something.

"One more question, if I may. Do you feel you can remain friendly and sincere with Annelies?"

"Of course."

"Forever."

"Why, Doctor?"

"Have pity on this child. She cannot face violence or harshness. She dreams of someone who will love her, who will give her pure love. She feels like she is living alone, by herself, without knowing the world. She has put all her hopes for the future in you, Mr. Minke."

He was no doubt exaggerating.

"She has a mother who guides her, educates her, who loves her," I said.

"She doesn't fully believe that her mother's love will last. She waits in anticipation for that moment when her mother will explode and reject her."

"Mama is a very wise woman, Doctor."

"That cannot be denied. But Annelies cannot convince herself of that. Possibly, secretly, Annelies sees her mother as being more attached to the business than to her. This is just between the two of us. No one else need know of this conversation. You understand."

He was silent again for quite a while. Suddenly:

"So you understand?"

"I think so."

"There must not be any hard, harsh, disappointing words. She loves you. I speak to you like this, first of all, because Native men are not used to treating their women gently and politely, as friends and with sincerity, at least as far as I know, according to what I've heard and read. You have studied European civilization, so you no doubt know the difference between the attitudes of European and Native men towards women. If you are the same as most other Natives, this child will not live long. Quite frankly she could fall into a living death. If it came about, if, I say, you married her, would you take a second woman at some later stage?"

"Marry her?"

"Yes, that is what she dreams of. That you will marry her, yes? You're in your last year at school, aren't you?"

"I've no desire to propose yet, Doctor."

"If necessary I will propose on your behalf in order to save this girl."

I couldn't say anything.

"So you will marry her and will not take a second woman." He put out his hand to obtain the certainty of a promise from my lips.

I took his hand. I had never intended to take more than one wife. I always remembered the words of my grandmother: Every man who takes more than one wife is a liar, and will certainly become a liar whether he wants to or not.

"Her heart is too soft, too gentle. She can't cope with hurt. You must always humor, caress, protect her. It seems her self has been taken from her."

"Taken from her?"

"By someone very close to her."

"Who, Doctor?"

"I don't know. You will find out for yourself. Something around here, for certain. There are many secret, suppressed problems within her young heart. She lives, in fact, as an orphan. She

203

feels permanently dependent. Even here in her own private world, there is no one she can rely on. She needs someone who will support her. Having grown up in the middle of wealth, she doesn't appreciate the security it gives. For her, wealth is nothing. That is what I can understand about this child. You're listening, yes?"

Dr. Martinet took his monocle out of his top pocket and put it in his right eye. After looking at his watch, he stared at me.

"Thank you for listening so earnestly. Look at the calm and peaceful view from here. It is lucky that this girl lives in the middle of luxury and peace. I don't know what would happen otherwise."

The palakia tree seed in my head was replaced by another type of seed: suspicion. What did the doctor really mean?

"Excuse me. I'm not a psychologist. I've spoken a lot with her mother—an amazing woman. She is a mature and civilized person. She has real strength of character, reinforced by the hardness of someone with revenge still in her heart. Just for her, as a woman, to be so educated, is extraordinary. Even in Europe it would be an astounding feat. I don't think she has developed in this way consciously. One or indeed many experiences have been the motor behind these changes. I don't know what they might have been. Her heart is very hard, her mind very sharp, but in all this it is her own success in all her endeavors that has made her into such a strong individual, and so daring. But she has one big failure in a certain matter. It's understandable: Every self-educated person has a failure that stands out."

Dr. Martinet didn't continue. He hoped that I would seek for myself the meaning of his words.

"She's beginning to regain consciousness, your Annelies," he said all of a sudden. He looked over towards her, left me, and approached his patient. He checked her pulse, then he waved to me. "Yes, Mr. Minke. In a few minutes she will return to being the Annelies you knew before. May she return to complete health now that you are here again. From this moment on, this girl is no longer my patient but yours. Everything I have told you is between us alone. Good afternoon."

He left the room, closed the door behind him, and disappeared from sight.

Now the moment had arrived when I could feel sorry for myself. Yes! One after the other over these last few days new experiences had fallen upon and rocked me. And now there was a new one that I had to face: Annelies.

Great artists, Minke, said Jean Marais once, whether they are painters or something else, or leaders, or generals, become great because their life has been crammed with and based upon profound, intense experiences: emotional, spiritual, or physical. He had said this after I had finished telling him the life story of the Dutch poets Vondel and Multatuli. Without such profound experiences, greatness is purely imaginary: Their greatness is whistled up by the money-minded people of the world.

Jean Marais did not know that my own writings had begun to be published. If his words are true, I thought, maybe one day I could become a great writer like Hugo, as Nyai hoped. Or a leader, or a teacher of a nation as hoped by the de la Croix family. Or perhaps I will end up as rotting flesh just as Robert Mellema (if Darsam's story was true) and Fatso planned.

Annelies sighed and moved her finger. She will be all right; I will not have to watch her die. I moved away and sat on a chair where I could watch her. Even ill she was gloriously beautiful: Her skin was fine, her nose, eyebrows, lips, teeth, ears, hair . . . everything. And I began to doubt Dr. Martinet's explanation of Annelies's psychology. Could such a beautiful body house such a disordered mind? And I—an outsider, just an acquaintance—must I also accept some responsibility for her just because of her beauty? Creole beauty. How involved my life was becoming! The result of my own actions as a philogynist.

"Mama!" uttered Annelies. Now her legs began to move. "Ann!"

She opened her eyes and continued to gaze off into the faraway distance. From this moment on she was my patient; that's what Dr. Martinet had said. I held back my laughter, understanding that I was now the doctor who must cure her.

I took the milk from the table. I raised her head with my arm and poured a little milk into her mouth. She began to sip it and smacked her lips a little. Yes, she was beginning to regain consciousness. I gave her some more to drink. She began to swallow.

"Ann, my Annelies, drink it all up," I said and I gave her some more to drink.

She took one swallow after another.

Nyai entered carrying lunch for two people.

"Why are you doing it yourself, Mama?"

"It's not that. No one else is allowed to come up here. So the doctor was right—she is beginning to wake up now."

"Almost, Mama."

"Yes, Minke, the doctor said only you can look after her now. It's up to you," and she went out again.

Annelies opened her eyes again and began to look at me.

"What's wrong with you, Ann?"

She didn't answer, but just gazed at me. I put her head down on the pillow again. The beautiful shape of her nose pulled my hand over to stroke it. The ends of her hair were brown and her eyebrows were lush, as if they had been fertilized before she was born. And her eyelashes, so long and curly, made her eyes seem like a pair of morning stars in a clear sky, her countenance itself the clearest sky of all.

Where else on this earth of mankind could one discover such perfect Creole beauty, such a beautifully harmonious form? God has created such a thing only once and only in this one body in front of me. I will never let go of you, Ann, whatever is going on inside you. I am ready to face whatever and whomever.

"Ann, today," I said to her, "the weather is beautiful. It is hotter than usual, but it's fresh, not too humid."

The girl still just looked at me, her gaze focused on the tip of my nose. She still hadn't spoken. Her eyes blinked so slowly. Yet her beauty was still profound, greater than all those things that have been made by man, richer than all the combined and individual meanings to be found in the treasuries of the languages. She was a gift from Allah, without equal, unique. And she was mine alone.

"Arise and awaken, Flower of Surabaya! Do you not know? Alexander the Great, Napoleon, all would fall to their knees to gain your love. To touch your skin they would sacrifice their nations, their people. Awaken, My Flower, because the world is a lesser place without you," and without knowing it I was kissing her on the lips, and then became fully conscious of what I was doing.

The long breath she expelled blew over my face. Her lips smiled. Her eyes too. But she still could not speak. So I kept on with my chatter, like Solomon praising the virgins of Israel: chin, breasts, cheeks, legs, the look in her eyes, her eyes themselves, neck hair—everything, all of her. I stopped only after I heard:

"Mas!"

"Ann, my Annelies!" I said, cutting her off, "you're better now. Come on, get up. Let's walk. Come on, my goddess."

She began to move. Her hand waved to me. And I responded to that hand.

"Let me carry you," and I carried her. Yes, I carried her. And I wasn't strong enough. What sort of body was this, incapable of even carrying a girl! I put her down. Her legs stepped forward, shaking; her body swayed. I supported her. To the devil with chairs, table, and bed. I took her to the window where a minute ago I had stood with Dr. Martinet and where he had appointed me her doctor. A vast panorama of fields opened up before us. And the sun had already begun to leave its midday position.

"Look over there, Ann, the forest is the limit of our view. And the mountains, and the sky, and the earth. You see, Ann? Do you really see it all?"

She nodded. The wind whistled as it launched a gusty attack from out of that great expanse of nature, and it felt as if it was being channeled into that one window. Annelies shuddered.

"Are you cold, Ann?"

"No."

"You should get some more sleep."

"I want to be close to you, like now, Mas. It was such a long time, and you still didn't come."

"I'm here now, Ann."

"Don't let go of me, Mas."

"You're cold standing here."

"I'm warm enough now. The forest seems different some-how. And the wind too. And the mountains. The birds also."

"You're well now, Ann. You're beginning to be strong again."

"I don't want to be ill. I'm not ill. I was only waiting for you to come."

I no longer felt ill either, Ann, had you wanted to know. Something made me turn and through a small opening in the door I saw Nyai and Dr. Martinet. They didn't enter. Then the door closed again.

13

The school director excused my absence, which went over the period allowed in my doctor's certificate. The greetings I passed on from Herbert de la Croix softened his attitude. For several days I worked hard to catch up on my studies. It was easy. My grandfather had taught me that if you believe you will be successful in all your studies, then you will be successful; if you think of all study as easy, all will be easy. Be afraid of no kind of study, because such fear is the original ignorance that will make you ignorant of everything.

I followed his advice; I believed in the truth of his wise words. I never fell behind in my studies, even though, yes, even though I did not study as hard as the others. But now I really studied hard, to catch up on what I had missed during the last few weeks.

A carriage and driver had been put aside by Mama especially for me. Night and day. And each time I left for school in my new vehicle I picked up May Marais to drop her off at her school in Simpang.

Everything had changed. Especially, and most of all, myself. I now felt like a man of real substance as I sat in my luxury buggy in the middle of Surabaya's traffic. It was easy to see too that my friends at school had also changed. Meaning: They seemed to be, and probably indeed were, distancing themselves from me. I interpreted all this as a sign of respect towards someone who has just scored a rise in his marks. It was possible I might have been wrong in this estimation of my position, so I looked upon it as provisional only. My teachers, now that I traveled by luxury carriage, seemed to treat me as someone they didn't know but who was of equal status. This too was a provisional guess.

I felt I was no longer the old Minke. My body was the same, but its contents and its perceptions were new. I no longer liked to joke. There was more to me now than that. I was more thoughtful, while my friends at school were still childish. I no longer wanted simply to float on the surface of problems—in every conversation and discussion I wanted to dive straight down to the bottom of every problem.

See, even Robert Suurhof still didn't want to approach me. He always moved away if we passed near each other. And the girls at the school avoided me too, as if I were the source of some plague.

Several times the school director summoned me to get assurances that I had not already married, because a student must leave the school once he is married. I think it was Suurhof who was doing the talking. It could have been no one else. He alone knew what had happened. Eventually I found out for certain that my guess was not wrong. He'd spread rumors, inciting my friends against me. (So my estimation of myself was wrong after all!). The looks directed at me were from people I felt I no longer knew.

Everything had changed. Now, all around me at school, there was no enveloping aura of brightness, but only loneliness, a loneliness that called and summoned me to reflection.

The only teacher who did not change was Miss Magda Peters, the Dutch language and literature teacher. She still hadn't married. All over her exposed skin there were brown freckles. Her clear brown eyes were always sparkling. At first her appearance tended to make you laugh. She struck me as looking like a white, female monkey with an ever-surprised face. But then as we listened to her first lesson, we all became quiet. The impression of a white, female monkey disappeared. Her freckles vanished. A feeling of respect replaced all this. And here are her words, when, in her first class since traveling down from the Netherlands to the Indies, she said:

"Good afternoon, students of H.B.S. Surabaya. My name is Magda Peters, your new teacher for Dutch language and literature. Please put up your hands if you don't like literature."

Almost everyone put up their hands. There were even some who stood up to show their antipathy.

"Excellent. Thank you. Please sit down everybody. Even the people of the most primitive society—in the heart of Africa, for example—who have never sat in school, never seen a book in their life, who don't know how to read and write, are still able to love literature, even if only oral literature. Isn't it an outstanding achievement that after at least ten years in school, H.B.S. students still do not like literature and language? Yes, it's truly outstanding."

No one laughed and there was nothing to laugh at. Total silence.

"You will all advance through school. Perhaps you will obtain a string of all sorts of degrees, but without a love of literature, you'll remain just a lot of clever animals. Most of you have never seen the Netherlands. I was born and brought up there. So I know that every Hollander loves and reads Dutch literature. People love and honor the paintings of van Gogh, Rembrandt—our own and the world's great painters. They who do not love and honor them and who do not learn to love and honor them are considered to be uncivilized. Painting is literature in colors. Literature is painting in language. Put up your hand if you don't understand."

To make sure we weren't classified as uncivilized, from that moment on we all knew we would have to concentrate on the teacher's every word. She had us in the palm of her hand.

And Miss Magda Peters's attitude towards me never changed. She must surely have heard the rumors whispered by Robert Suurhof.

It was generally Magda Peters who opened the Saturday afternoon discussions. She did this not just happily but with great enthusiasm. Every student could put forward any topic—general or personal, local news or international developments—as the afternoon's subject. If no student had anything to put forward, only then would the teacher choose the subject. Students were not obliged to attend if they weren't interested. The most popular discussions were those led by Magda Peters. No one wanted to miss out, and so the discussion had to be held in the hall with the students sitting on the floor. Only the speaker would stand. The teachers also sat on the floor. The teacher leading the discussion would also stand. On those occasions, one could see that freckles covered all of Magda Peters's body.

In order to place my situation and my attitudes in context, so that you may judge the truth or otherwise of my views about myself and my surroundings, I think it is proper that I tell something about my experiences in those discussions.

Once I asked about the Association Theory of Dr. Snouck Hurgronje. Magda Peters asked for comments from the other students. No one knew anything about it. She then looked politely at the teachers but none showed signs of wanting to respond. Then she herself spoke.

"I am not sure what it means, either. Perhaps it is something that has arisen in colonial political life. Do you know what colonial politics are?" No one answered. "It is a system or power structure to consolidate hegemony over occupied countries and peoples. Someone who agrees with such a system is a colonialist, who not only agrees with it, but also legitimatizes it, carries it out and defends it. The basic issue in all this is the one of earning a living. None of this need yet attract your attention, students. You're all still too young. If such matters were to become the subject of a literary work, it would be much more interesting, as in the case of Multatuli's works, which have been discussed with you in class several times. Try, Minke, to explain Dr. Snouck Hurgronje's Association Theory."

I explained what I'd heard from Miriam de la Croix and my own views on it.

"Stop!" said Magda Peters. "Such subjects may not yet be discussed at H.B.S. school. It's up to you if you want to discuss them outside school. Such matters are the affair of the Queen, the Netherlands government, the governor-general and the Netherlands Indies government. If you have a desire to find out more, it's best you do so outside school. Because none of you have a topic for today's discussion, I will choose one myself.

"Just recently I came across an article about life in the Indies. Too few people write about this. Precisely because of that, it attracted my attention. Maybe the writer is Indo-European.

Maybe, I say. Perhaps some of you may already have read it? It's called: "Uit het Schoone Leven van een Mooie Boerin—The Beautiful Life of a Beautiful Peasant Girl." The writer's name is Max Tollenaar."

Several hands shot up. I kept a straight face. Max Tollenaar was my pen name. The original title had been changed and the editor had made some alterations to the text, not all of which I agreed with.

Miss Magda Peters began to read it out, placing stresses and pauses in such a way that her voice sang and the essay sounded more beautiful than I had originally intended. Yes, you could say it sounded like a long poem, dense with emotion. Virtually no one even blinked while listening. And after the reading was over and people were freed from its hold over them, at last they were able to let go of their breath.

"It's a pity that this story was published in the Indies, about the Indies, about the people and society of the Indies. As a result no one has discussed it in class. All right, one of you come forward and give us your reactions and comments on this story, perhaps even a critique."

Robert Suurhof immediately came forward. He stood there, legs apart, feet nailed to the floor, as if he were scared the wind would blow him over. All eyes were directed at him. Only I hesitated.

He looked around at his friends first. Perhaps he was looking for moral support.

"I've read four pieces by Max Tollenaar over recent weeks. They have all been about the same thing and have been colored by the same emotions. The writer is in the power of some force outside himself. Yes, yes, the writer is suffering some drastic fever. The writings are the long deliriums of someone who has lost all control over himself, who has lost all touch with reality. I don't know who Max Tollenaar is. But I can make a good guess, because I am the only witness to the events that he writes about.

"Miss Magda Peters, I don't think we should be talking about such writings in a school discussion," he continued. "It only makes us all dirty, miss. If I'm not mistaken—and I'm certain I'm not—the writer of this doesn't even have a family name."

He was quiet for a moment, glancing around at all the students in whom he was building up such tension. He raised his chin. His eyes shone victoriously. He was going to let go one more shot.

Miss Magda Peters looked taken aback. Her eyes were blinking rapidly.

I alone knew what Robert Suurhof was up to: to revenge himself upon me. So I began to understand better: It was he who wanted Annelies. There could be no reason other than jealousy for his hatred and his public insults to me. Yes, it was Robert who desired to possess Annelies. He took me with him that day to make himself look good, and so I could be a witness. Why me? Because I was a Native. Upper-class European ladies used to take a monkey with them everywhere so they would appear more beautiful in comparison. He took me. It was Suurhof's monkey that won Annelies's heart.

"The person in question, miss," Suurhof resumed, "is not even an Indo. He is lower than an Indo, than someone whose father refused to acknowledge him. He is an *Inlander*, a Native who has smuggled himself in through the cracks of European civilization."

He bowed to pay respect to Magda Peters and then to the other teachers present, and then sat down anxiously on the floor.

"Students, Robert Suurhof has just given us his opinion on the author of this story; an author whose identity he alone knows. What I had hoped for was an opinion about the story itself. Very well. Who do you reckon is the author?"

All the students looked around at each other. Then all looks were directed towards their friends who were neither Pure nor those who looked Mixed-Blood, but towards those who looked Native, as if underlining Suurhof's words. They all bowed down their heads.

I knew Suurhof's face was pointed towards me. The others followed his example. No, said my heart, no, don't be afraid. To the devil with all this; if need be, I can leave this school. If necessary, right now.

Suurhof stood up again. He said briefly:

"The author is among us now."

It seemed his whisperings had spread throughout the school. Now every face was directed towards me. I looked straight at Suurhof. Victory shone from his eyes.

"Who is it, Suurhof?" asked Miss Magda Peters.

With Caesar's forefinger, he pointed at me.

"Minke!"

Magda Peters took a handkerchief from her bag and wiped her neck, then her two hands. She didn't know what to do. She turned for a moment towards the row of seated teachers, then at me, then at the students seated upon the floor. Then she walked over to the teachers and the director, who also happened to be present. She gave a little nod, returned to the middle of the meeting, and made a path through the students, heading straight towards me.

Now I will be expelled and publicly humiliated.

She stood for a second before me. The freckles on her legs became clearly visible. And I heard her call:

"Minke!"

"Yes, miss." I stood up.

"Is it true you wrote this"—she held up the *S.N. v/d D*—"using the pen name of Max Tollenaar?"

"Have I done something wrong, miss?"

"Max Tollenaar!" she whispered and held out her hand. "Come," and she pulled me up and took me to the director.

All eyes were directed my way. Standing there before all the teachers and the director, I nodded respectfully. They hardly responded. Then I was taken across to stand before all the students.

Silence.

The woman teacher rested her hand on my shoulder. I was like someone at confession who didn't really know what to confess.

"Students, teachers, Director, today I introduce to you all, especially to you students, an H.B.S. pupil by the name of Minke, whom no doubt you all know. But I'm not introducing the Minke whom everybody knows, but rather a Minke of a different quality, a Minke whose use of Dutch to state his feelings and thoughts is brilliant, a Minke who has written a literary work. He has proven that he is capable of writing perfectly in a language that is not his mother tongue. His has brought to life a snippet of reality, which other people, even though they too have experienced that reality, could never explain. I'm proud to have a pupil like him."

She shook hands with me. I still wasn't told to go. Her words of praise raised me up to the highest of heights. Now I waited for the final chop to fall.

"Minke! Is it true you do not have a family name?"

"Yes, miss."

"Students, having a family name is just a custom. Before Napoleon Bonaparte appeared on the stage of European history, not even our ancestors—not one of them—used family names." She began to tell how Napoleon's decision on this was made law in all the territories that he controlled. Those who could not find an appropriate name were given one at the whim of the local officials, and Jews were given the names of animals. "Even so, the use of family names is not unique to Europe or to Napoleon, who got the idea from other peoples. Long before Europe was civilized, the Jews and Chinese were using clan names. It was through contact with other peoples that Europeans learned the importance of family names." She stopped.

I was still standing there for everybody to look at.

"Is it true you're not an Indo, Minke?"—a formal question that I had to answer in the affirmative.

"*Inlander*, miss—Native."

"Yes," she said loudly. "Europeans who feel themselves to be a hundred percent pure do not really know how much Asian blood flows in their veins. From your study of history, you will all know that hundreds of years ago, many different Asian armies attacked Europe, and left descendants—Arabs, Turkish, Mongol—and this was after Rome had become Christian! And don't any of you forget that under the Roman empire, the Asian blood, and perhaps even African, of those citizens of Rome from various Asian nations—Arabs, Jews, Syrians, Egyptians—now mingles with the blood of Europeans."

Silence continued to reign supreme.

My heart was empty. My body seemed unsteady. My only desire was to sit down again.

"Much of Europe's science comes from Asia. Yes, even the numerals you use each day are Arab numerals, including zero. Imagine, what would it be like to count and add up without Arab numerals and without zero? And zero in its turn was derived from Indian philosophy. Do you know what the meaning of philosophy is? Yes, another time we'll talk about this. Zero, a condition of emptiness. From emptiness comes the beginning, from the beginning there is a development until the climax, number nine, and then there is emptiness and we begin again with a higher value, tens, and so on, hundreds, thousands . . . there is no limit. Without zero the decimal system would vanish, and all of you would have to count using roman numerals. Most of your names are Asian, because Christianity was born in Asia."

The students began to show their restlessness.

"If Natives do not have a family name, that is because they don't, or don't yet, need one; and there is no humiliation in that. If the Netherlands doesn't have a Prambanan or a Borobudur temple, it means in that era Java was more advanced than the Netherlands. If the Netherlands still does not possess such things, yes, it is because they have never been needed . . ."

"Miss Magda Peters," the director intervened, "it's best that this discussion be closed."

The discussion closed; everybody dispersed. Except for Magda Peters everyone seemed to avoid me. No one called out as usual. No one laughed. No one raced ahead of each other as they usually did. They all walked off quietly, full of thought.

Jan Dapperste, a student whose appearance was more Native than European, stood at the fence following after me with his eyes. He always introduced himself as Indo. But to me alone he had admitted to being Native. Trusting in me as a friend, he had explained that he was the adopted child of a preacher named Dapperste. An adopted child! He himself was pure Native. He felt close to me. After I obtained the buggy, it was usual for him to ask to ride along with me. Now he too seemed to be keeping his distance.

This time it was Magda Peters who asked for a ride. She didn't say a word the whole way. Indeed, what's the use of speaking when your heart and mind are full of troubles? The traffic was invisible to me. I could see only one thing: the students' and the teachers' anger towards Magda Peters. Their Europeanness had been wounded.

Once or twice Magda Peters looked at me from beside me where she was sitting.

"What a pity," she sighed into the wind.

I pretended not to hear.

The buggy stopped in front of her house. She said thank you. Then suddenly:

"Come in, Minke," and that was the first time she invited me into her home.

I walked with her inside. So we sat facing each other on the settee in the main room.

216

"You're extraordinary, Minke. So you really wrote that."

"It's so, miss."

"You are certainly my most successful student. I've taught Dutch language and literature for five years now. Almost four years in the Netherlands. None of my students could write as well as that—and to be published as well. You must be fond of me—are you?"

"There is no teacher of whom I'm more fond."

"Is that true, Minke?"

"With all my heart, miss."

"I guessed so. You must have been following all my lessons very carefully, with all your mind and heart. Otherwise there is no way you could write as well as that. You're not angry with Suurhof, are you?"

"No, miss."

"Good. You're worth much more than he. You've proven what you can do."

The flattery was so embarrassing. She told me to stand up.

"At the very least, Minke, my efforts, my strivings these five years have now achieved some results." She pulled me near her.

Totally surprised, I found myself in her embrace, and she kissed me until I was out of breath! Until out of breath!

Every day I had to visit Jean's house, even if only for one or two minutes, to drop off or pick up May or to hand in some new order. I also had to drop in at my boardinghouse.

Having my own buggy made everything easier: chasing after orders, writing advertising texts, writing other things as well. My time somehow seemed to last longer.

When I arrived home, I was usually exhausted and needed to sleep for a while. Usually Annelies woke me up, bringing a fresh towel, and ordering me to bathe. Afterwards we sat and talked, or read Indies newspapers, or Dutch magazines.

At night, I worked, studied, or wrote while waiting upon Annelies in her room. Her health was improving with every day. But she hadn't yet resumed working.

Mama was very busy in the office and out in back; she had no time for the two of us during the day.

That night, like the nights before, I sat at the table in Annelies's room. She was reading Defoe's *Robinson Crusoe* in a Dutch

217

translation, each page of which was divided into two columns. I'd prepared a list of books she had to read, all books for young people, such as Stevenson and Dumas. She had to finish them in one month. And beside her lay the old dictionary that Mama used every day—an old dictionary that, over the last ten years, had become incapable of meeting the demands of new developments.

I sat across from Annelies reading letters from Miriam and Sarah before I started writing a story to be titled "A Father's Son." I meant no other than Robert Mellema.

This time Miriam's letter was even more splendid.

Do you remember at all that "other one"? I've received a letter from the Netherlands. From a friend, a close friend, who knows what happened to him in South Africa, in the Transvaal. The writer of the letter returned home to the Netherlands after being wounded in a brief battle. He himself was once in the same unit as the "other one." The brigade was under the command of one Mellema, a young engineer who was hard, courageous, ambitious, he said.

Friend, I was so happy to receive his letter. As happy as I am to receive yours. In it, friend, there was something mentioned that might interest you. The "other one" is a few years older than you, perhaps. Answering the Dutch call to seize back and defend their independence from the British, and without thinking too much about it all, he left for Africa . . . and was greatly disappointed.

Though there are some reports on the war in the Indies press, there is a great deal that is not reported. The Dutch were immigrants there, my friend—I think your favorite teacher, Magda Peters, pays too little attention to wars—and they ruled over the native peoples. In their turn the Dutch immigrants were conquered by British power, also an immigrant power from Europe. So power was structured in layers, with the natives at the bottom.

Just think, isn't it the same in the Indies? Just as Papa explained? There are some small differences, but they don't alter the basic reality. Aren't the Natives here ruled by their own rulers? Kings, sultans, and bupatis? In their

218

turn this brown government is controlled by a white government. The kings, sultans, and bupatis, with all the facilities they have here, are the same as the immigrant Dutch power in South Africa.

My friend, that "other one" was so disappointed when he realized what the war between the British and the Boers—the Dutch immigrants—was really about: who would control the land, gold, and natives. The young Dutchmen who were called to go there from all over the world came only to be wounded or to die for interests alien to Holland. That "other one," according to his letter, saw how the natives of South Africa were much worse off than the natives of the Indies, far worse off than the natives of Aceh. If he was honest to himself, he said, he felt no different from an Indies Army soldier in Aceh.

But he realized this all too late. And he only came to that realization as a result of an unexpected meeting with a nonwhite inhabitant, though not a black, named Mard Wongs. This person, my friend, was only one of several wealthy farmers who could speak Javanese. He and all the others, even though they spoke Afrikaans, were Slameiers, of your own people. Mard Wongs was an Afrikaan version of his original name. I think it must have been: Mardi Wongso. And the Slameiers are none other than the descendants of Javanese and Buginese-Makassarese-Madurese Natives, who had been exiled to South Africa by the Company.

Interesting, yes?

Now Mellema's platoon, so writes my friend from the Netherlands, entered Mard Wongs's house to shelter there for the night. The old man, who was white with age, refused them and angrily threw them out of his house. Mellema lost his temper and threatened the old man that he would be shot.

Mard Wongs became more enraged: What else do you Dutch want? In Java you robbed us of all that we rightfully owned, you robbed us of our freedom, and now here you beg for shelter under my roof. Have you never been taught the meaning of robbery and begging?

Shoot me then! Here is the breast of Mard Wongs. I would not give you the shade from a single piece of roof nor shelter behind one board of my house. Go!

And do you know, my friend, that in this contest of wills, Mellema gave in. He and his platoon were forced to sleep out under the open sky.

It was that incident that made the "other one" realize how Indies Natives hated the Dutch. He realized then that he and his platoon were not the upholders of noble ideals, but merely and no more than the tools of colonial power. He was ashamed. He was confused. He had dreamed of becoming a hero, of making some contribution to humanity. Now he was in the middle of tyranny's arena.

Pity the "other one."

The next morning his platoon attacked a position held by the British South African Light Horse Brigade. Earlier, a Boer regiment had attacked in large numbers from another direction but was confronted, pushed back, and almost totally surrounded and annihilated.

In the midst of all this, Mellema's company attacked the enemy. The British were surprised, fell into disarray, and dispersed under alternative attacks from two directions. The position fell to the Boers.

But, my friend, "the other one," was shot and captured. He has written that perhaps he will be taken as a prisoner of war to England. During those last days he never tired of regretting his earlier stupidity.

The reason why I'm telling you all this, my friend, is simply to add another perspective about something that is not generally publicized in the Indies. Isn't it true you only read about the brutality of the British and the victories of the Dutch in the papers? On the other hand, says Papa, the British press reports are about the savagery and viciousness of the Dutch towards the natives. But there is no paper in England, the Netherlands, or the Indies that talks about the South African natives themselves, let alone about the Slameiers people. Isn't the world strange?

I think the Javanese Natives are better off. There

have been a few people who have spoken up on their behalf. Yes, even though their voices have been drowned in the sheer din of the bureaucracy. This is something we haven't talked about and analyzed yet. Let's try to discuss it another time. You agree, I hope?

Now Minke, my friend, don't let me wait so long for your next letter.

Miriam de la Croix.

Sarah's letter was different again. She wrote:

I can understand it if Miss Magda Peters didn't know anything about the Association Theory. We didn't know anything more than we told you that day. No more than that.

I told Papa that you didn't know anything about it. He only laughed boisterously, and said: You also do not know any more than the little you told him.

After your letter arrived I told Papa that Magda Peters didn't seem to know anything about it. Your other teachers couldn't offer an explanation either. Perhaps they were unwilling, deliberately stopping themselves from answering, or perhaps, indeed, they didn't know anything. So what did Papa say? Not everyone has an interest in colonial policy, just as not everyone is interested in the art of cooking. And don't forget too, that in this age in which we live all the Indies believes in the greatness, authority, wisdom, justice, and compassion of the government. There are no beggars dying of hunger in the streets. Neither are there sick people dying in the streets. They too are protected by the government's laws. No foreigner is beaten to death, just because he is a foreigner; foreigners too are protected by the government's laws.

There is something I feel you should know. Papa has been talking about you: A youth like that should continue his studies at a university in the Netherlands. Perhaps he should study law, Papa said of you. Even if he failed, he would still have learned what the law means to Europeans.

What do you think? Is it possible that a Native could become a graduate in a European science? To be honest, Papa doubts it. Papa says—don't be angry like you were before—the Native psychology hasn't yet developed as far as that of the European: His wiser considerations are still too easily pushed aside by lustful passions. I don't know if this is true or not. It seems that it is true though, especially if you look at the upper echelons of your people. You too should think about this. What do you think?

There is another thing I should pass on to you: One of the youths being tried out by Dr. Snouck Hurgronje is called Achmad, from Banten. I tell you this in case some day you meet, become acquainted, and correspond.

"Why are you sighing?" Annelies suddenly asked.

"On fire."

"What's on fire?"

"My head. My own head. All sorts of things keep coming up. There is already so much work and I'm still not allowed to go undisturbed for a single moment. Read!" and I pushed the letters over to her.

"They're not for me, Mas."

"It's best you know too."

Annelies read them slowly and carefully.

"It seems many people are fond of you. It's a pity I don't understand much of this."

"There's no reason to say it's a pity, Ann. They all seem to want to be my teacher."

"Isn't it good to have teachers?"

"You too, Ann! Of course, it's good to find a teacher. No knowledge is useless. It's only that they all seem to have a passion to see me become an important person as a result of their own efforts. Aren't they capable of doing it themselves anyway? Boring teachers are a terrible torment, Ann," I said.

"Then you don't need to answer."

"That's not right either, Ann. I've read their letters. They wrote to get an answer."

And Sarah had gone a bit too far. Unashamedly mentioning things like lust. Asking for a reply too. Does she want me to strip myself naked? Even in Europe this is not yet a matter for public discussion. It is a private, tightly closed matter. These de la Croix girls go too far!

Annelies resumed reading. I think it unsettled her that the letters came from two girls, sisters. She put the letters down on the table, folded them properly, and put them back into their envelopes. She didn't say anything else.

Neither of us spoke for some time.

"Ann," I began, "you seem to be getting better."

"Thank you for your treatment, Mas Doctor."

"From tomorrow you no longer need a friend to stay in your room with you."

She looked at me suspiciously.

"You're not going back to Kranggan?"

"If you still want me to stay, then, of course, I won't go back."

She frowned. For a moment, she glanced at the letters from Sarah and Miriam.

"Don't you want to stay with me anymore?" she asked in a voice that sounded as if she was about to cry.

"Of course, Ann, while you're ill."

"Must I become ill again?"

"Ann, what are you saying?" and at that moment I remembered Dr. Martinet's warning. And I was sure I hadn't spoken roughly to her. I quickly added: "You must recover fully, Mama needs you greatly!"

"Why don't you want to stay here with me just because I'm not sick anymore?" she asked nervously.

"What will people say?"

"What will they say, Mas?"

"Look, Ann, let me explain it to you: You're better now. If you don't want me to go back to Kranggan, then I won't. Believe me. I'll stay here at Wonokromo as long as you want. But not, of course, in your room. So beginning tomorrow, Ann, I want to stay and work in my own room, near the garden. If you feel lonely, you can come and visit me. It's just the same, isn't it?"

"If it's just the same, let's just keep on like this forever. You stay here in my room."

"But upstairs is out of bounds for everyone except Mama and you. We must respect the rules, mustn't we?" and there were still another twenty sentences that I uttered.

223

She didn't interrupt any more. But her stare reached farther and farther out into the distance. Annelies was jealous.

The next day I visited Jean Marais. Before I left home, I prepared a question about South Africa. He listened silently.

"You know, Minke, as a European I'm already very ashamed that I have become involved in colonial affairs. Probably I'm very similar to the person you've just described, someone whom neither of us has met. I fought in Aceh only because I assumed Natives would not be able to fight back, and so wouldn't fight back. But they fought back all right, really fought, with all their might, ignoring all obstacles. And they were courageous and daring too. Just as in many of the great wars of Europe. It was something to be very ashamed of, Minke: Europe's latest weapons were pitted against the flesh of the Acehnese. Because you've asked me my opinion I'll answer, but never ask me about these things again—it tortures my conscience."

Without us realizing, Mr. Telinga had begun to listen in from a distance; then he came closer, sat upon the table. It looked as if he was eager to join in the discussion.

"All the colonial wars for the last twenty-five years have been fought in the interests of capital; fought to ensure markets that would guarantee more profits for European capital. Capital has become very powerful, all-powerful. Capital decides the fate of humanity."

"War has always been the clash of force and of strategies so as to emerge victor," Telinga intervened.

"No, Mr. Telinga," Marais protested, "there has never been a war conducted for its own sake. There are many peoples who go to war who have no desire to be victor. They go to war and die in thousands, like the Acehnese now . . . because there is something they want to defend, something more important than death, life, or defeat and victory."

"It's all the same in the end, Jean. A contest between force and strategies to emerge victor."

"That's only how it ends up, Mr. Telinga. But if that's your opinion, very good. Now if, for example, Aceh wins and Holland is defeated, will the Netherlands become an Acehnese possession?"

"There is no way Aceh can win."

"Yes, that's precisely the point. The Acehnese themselves know they can't win, while the Dutch know too that victory will surely be theirs. Yet, Telinga, the Acehnese still descend to the battlefield. They don't fight to win. They're different from the Dutch. If the Dutch thought that Aceh was strong enough to defend itself, they would never dare attack, let alone start a war. The whole thing is a matter of calculating the profit and loss of capital. If the whole issue is just a matter of winning, why doesn't Holland attack Luxembourg or Belgium, since they're both closer and richer?"

"You're a Frenchman, Jean. You don't have any stake in the Indies."

"Perhaps. At the very least I regret ever having taken part in the war here."

"But you, like me, still accept the army pension!"

"Yes, just like you. But that pension is my right; due to me from those who sent me to war. Just like you. I lost my leg, you lost your health. That has been the only consequence of the war for both of us. We don't want to have an argument, do we, Telinga?"

"You never used to talk like that when you were in the platoon!" accused Telinga.

"Then I was your subordinate, now I'm not."

"What's the point of this argument?" I intervened. "I only asked about South Africa. Good-bye."

Then I went away and visited Magda Peters. She shook her head: "About South Africa? Do you want to become a politician?" she asked me in return.

"What is a politician really, miss?"

Once again she shook her head and looked at me as if I were someone suffering some grief. All we could do was sit there in silence.

"Later, when you have graduated. Then we can talk about this calmly. There is no need to talk about it now. You must make sure you graduate. Your marks are, indeed, not bad. With better ones you're sure to pass. Don't think about other things. Minke, are the stories true—I don't know where they come from—that you're living with some nyai or whatever?"

"Yes, miss."

"You know what people think of them?"

"I know, miss."

"Why are you doing it then?"

"Because it is not important where you reside. Especially when someone we call a nyai on the outside, miss, is actually no less than an educated person; indeed she is my teacher."

"Teacher? What does she teach you?"

"How someone can, starting from nothing, become an out-standing, self-educated person."

"Self-educated in what?"

"First of all in developing herself, then in developing a big business."

"Don't defend yourself with lies."

"I think I've never lied to you, miss."

"No, except this once." She looked at me, her eyes blinking rapidly, a sign (according to my guess) that she was thinking hard. "Don't disappoint me, Minke. You're educated. It's not fitting that you should act as if you'd never been to school."

"What I said just now was my answer, as an educated person."

The worry in her eyes began to go away. She blinked rapidly again, only it didn't seem funny anymore.

"Explain to me how a nyai can become a self-educated person. Educated in a European way is what you mean, isn't it?"

"At least as I understand it, miss. Perhaps I'm wrong; but try to visit us when you're not busy—in the evening, for example. Miss will be escorted home. They don't usually receive guests, but miss will be my guest."

"Very well," she said, accepting my challenge.

And I knew for certain that she would come.

"Would you like to visit now?"

"Very well. You should know, Minke, that I need to know what's really going on—to pass on to the Teachers' Council. Something could happen to you one day, Minke."

So we departed. We arrived at five o'clock in the afternoon. I took her into the front parlor and asked her to sit down. I studied the look on her face.

"It's nothing like I expected," she whispered. "In the Neth-erlands and Europe, a house like this . . . so this is where you

226

live?" I nodded. "It's not easy to own a house like this. Or to live in one. Oh, Minke, like the German houses of Central Europe."

Her attention was caught by something. I followed her gaze.

Annelies, in her black velvet gown, had entered the front room.

"Ann, this is my teacher, Miss Magda Peters."

Annelies approached, bowed, smiled, and held out her hand. And my teacher seemed bewitched. Her eyes had no chance to blink. She stood and shook hands, her mouth wide open.

"Annelies Mellema, miss. Just recovering from an illness. Ann, would you like to call Mama?"

Annelies nodded as she took leave, and left without saying a word.

"Like a queen, Minke. Her face is so delicate. Like an Italian prima donna. Is she the nyai's daughter?" I nodded. "She seems well educated, polite, and grand. Is it because of her you're staying here, Minke?" I didn't answer. She had to understand my silence. "It looks as if she might be your character in "Uit het Schoone Leven van een Mooie Boerin.""

"It's indeed her, miss."

"Prima donnas from Italy and Spain, ballerinas from Russia and France, none are as beautiful as she," she said as if she were lamenting her own fate. Then, as if addressing herself, "Yes, I can see why so many people talk so much about Creole beauty. It's a pity, though—the gown should be worn in the evening."

Mama arrived on the scene in her usual clothes: a white embroidered Javanese blouse worn with a brown, red and green kain. She held out her hand to the guest.

"This is Mama, miss, and this is my teacher, Mama. Miss Magda Peters, my Dutch language and literature teacher. Mama is not used to receiving guests, miss," I said, excusing myself to both parties, and because I had brought my teacher here without Nyai's consent.

It appeared that Mama was not upset by my audacity; rather, she began:

"Are Minke's studies going well, miss?"

"He could do better if he wanted to," she replied politely.

"We are not used to receiving guests, miss," said Mama in

227

flawless Dutch. "We're very happy that you have taken the trouble to visit."

"Ma'am, my visit is actually related to school affairs. We want to find out for certain whether Minke can study properly here."

"He leaves in the morning and returns in the afternoon. In the evening he reads, studies, or writes. I'm sorry, miss, I'm not used to being called ma'am and indeed I'm not a Mrs. It's not an appropriate way to refer to me, not my right. Call me Nyai as other people do, because that's what I am, miss."

Magda Peters blinked rapidly. I could sense she had been shaken by this request from the woman standing before her.

"There's no harm in being called ma'am. There's no insult involved, is there?"

"There's no harm. And there's no insult. It's only that it contradicts reality somewhat; it's also not in accord with the law. I've never had a husband. There has only ever been a master who owns me, my person." In her voice I could hear the bitterness of her life: sharp, a protest directed at humanity.

"Owns?"

"That's what has happened, miss. As a European woman you would no doubt shudder to hear about it."

I began to feel uncomfortable listening to this conversation. Mama was seeking compensation for her past wounds. An unhappy conversation, for the person listening and the person speaking.

"But slavery was abolished in the Indies almost thirty-five years ago, Nyai," Magda Peters responded.

"Yes, miss, as long as there are no reports about slavery. I have read somewhere that there is still slavery in many parts of the Indies."

"From the missionaries?"

"My situation is about the same."

Magda Peters was silent for quite a while. She blinked rapidly.

"Ma'am is not a slave, and is not like a slave either."

"Nyai, miss," Mama corrected her. "A slave can live in an emperor's palace, but still remain a slave."

"Why does Nyai feel herself to be a slave?"

This personal problem, which had been buried so long, now,

before this European woman, was seeking for a way out into the open, to protest, to complain, to condemn, to seek attention, to accuse, to charge, to judge—all at once. As I listened I became even more anxious. My thoughts were busy trying to find some excuse to slip quickly away. Meanwhile Nyai was opening the door to her past.

"A European, Pure European, bought me from my parents." Her voice was bitter and filled with a desire for revenge which would not be satisfied even with five palaces. "I was bought to become the brood mother of his children."

Magda Peters was silent. I hurriedly excused myself. I found Annelies upstairs by the window, reading a book.

"Why don't you come down, Ann?"

"I'm finishing this book."

"Why are you in such a hurry to finish it?"

"Really, I prefer hearing your own stories, Mas. Mas still hasn't told me very many. You get me to read other people's stories—these books. You do want to tell me a story, don't you?"

"Of course."

She resumed reading. All of a sudden she stopped, turned around towards me.

"Why have you come here? This is an out-of-bounds area, isn't it?"

"To call you down, Ann. Miss wants to talk with you."

She didn't reply and kept on reading. I went up close to her. I stroked her hair. She didn't react. When I pulled the book away from her, her eyes didn't follow my action. Annelies wasn't reading. She was hiding her face.

"What's the matter with you, Ann? Angry?" There was no answer. "It must be a good story you're reading."

She bent over and I could feel her shoulders shuddering as he held back her sobs. I turned her around to face me. All of a sudden she hugged me and burst out crying.

"What's the matter with you, Ann? I haven't hurt you in any way, have I?"

And I don't know if it was tens or hundreds of sentences I threw about everywhere to humor her. And she still didn't speak. She hugged me tightly as if afraid I would come loose and fly up into the green heavens. Annelies was jealous.

Conversation between two people drifted in through the

door, which wasn't properly shut. It became increasingly clear that it was originating from the upstairs corridor. Mama could be heard calling out. Annelies released me from her embrace. I poked my head out to have a look. Miss Magda Peters and Nyai were waiting for me in front of one of the upstairs rooms.

"Miss wants to see our library, Minke. Come, I'm going to show her." She opened the door to the room, a room I'd never entered.

The room was the library of Mr. Herman Mellema. It was as big as Annelies's room. Three big cabinets full of heavily bound books stood side by side. There was also a glass cabinet which held Herman Mellema's bamboo pipe collection. The furniture was all spotlessly clean. There was no carpet on the floor, exposing ordinary planks, neither parquetry nor polished wood. There was only one table with a straight-backed chair and an armchair. On the table there stood a white metal stand with fourteen candles. A book, which turned out to be a volume of magazines, lay open on the table.

"A very nice room, clean and quiet." Magda Peters allowed her gaze to wander around the room until it fell upon the glass windows that exposed views of the country outside. "So beautiful!" Then she went straight to the table and took up the volume of magazines. She asked, without looking at any particular one, "Who is reading the *Indische Gids*?"

"Bedside reading, miss."

"Bedside reading!" She stared wide-eyed at Nyai.

"The doctor advises that I read myself asleep."

"Nyai doesn't sleep well?"

"True."

"Nyai has had so many troubles for such a long time?"

"Five years now, miss."

"And Nyai hasn't fallen ill?"

Mama shook her head, smiling broadly.

"So what is Nyai looking for in these magazines?"

"Just something to make me sleep."

"What is Nyai's other bedtime reading?" she asked like a prosecutor.

"Whatever I grab hold of, miss. I don't really choose."

Magda Peters blinked rapidly again.

230

"And what does Nyai prefer most?"

"What I can understand, miss."

"Does Nyai know anything about Snouck Hurgronje's Association Theory?"

"Excuse me." Nyai took the magazine from my teacher's hands, looked for a specific place, then showed it to Magda Peters.

My teacher glanced over the page quickly, nodded, then looked at me.

"Why did you bring up the Association Theory at school? You should have just asked Nyai."

"I only wanted to find out more about it," I said, even though I never knew that this library existed or that there existed a magazine which published articles about such things.

Magda Peters then inspected the books in the cabinets. Most were beautifully bound volumes of magazines. It was as if she wanted to inspect the contents of Nyai's head. She didn't seem to find the collection very interesting: They were mostly about livestock, agriculture, commerce, forestry, and timber. But there were also volumes of general and women's magazines from the Indies, Netherlands, and Germany. Magda Peters's gaze swept quickly over most of the library. Then she returned again to the row of colonial magazine volumes and stopped for quite a while in front of a row of books of world literature, all in Dutch translation.

"There is no Dutch literature here, Nyai."

"My master wasn't very interested in Dutch literature, except for Flemish writings."

"Then Nyai must have read some Flemish books also?"

"Yes, there are some."

"May I ask why Mr. Mellema did not like Dutch literature?"

"I don't really know, miss. But he used to say that it was dominated by triviality, had no spirit, no fire."

Magda Peters swallowed. She didn't try to question Mama further. Then she shifted her attention again to the whole library, as if trying to give the impression that she had obtained a picture of the cultural level of Mama's family, a family much slandered at my school lately.

"May I speak with Annelies Mellema?"

"Ann, Annelies!" called out Mama.

I went to her room. I found her sitting by the window. Her gaze was occupied with the great broad panorama in the distance, the mountain range and forests.

"Don't you want to come, Ann?"

She still frowned. Didn't answer.

"Very well. Stay in your room, Ann." And I went and left her.

"Ann!" called Mama once more, softly.

"She's not feeling well. Forgive her, miss, she's just recovered from an illness."

The two women went downstairs to the garden veranda while busily chatting to each other. I don't know what about. One hour later I escorted my teacher home in the same buggy.

She invited me to come in and sit for a moment. But during the journey she did not speak at all.

"First, Minke, after seeing what that family is like, I feel that I'd very much like to visit there often. Your Mama is indeed extraordinary. Her clothes, her appearance, her attitudes. But there are too many sides of her personality. And, except for her language and the embroidery on her blouse, she is totally Native.

"That complicated personality of hers has come very close to taking on the progressive and bright aspects of Europe. And indeed she knows a great deal, too much for a Native to know, and a female Native at that. She is indeed fit to be your teacher. Only that growl of revenge in her voice and in the implications of her words . . . I couldn't bear to hear it. If that vengefulness was missing, she'd be truly, brilliantly outstanding, Minke. This is the first time I've met someone, and a woman too, who didn't want to make peace with her own fate." She let out a great long breath. "And it's amazing, her awareness of the law."

I was silent. There were several things I did not understand at all. I would ask Jean when I got a chance.

"Just like something out of a legend from *A Thousand and One Nights*. Imagine, she feels it's more proper that she be called Nyai. I thought it was only a part of her revenge. But, indeed, Nyai is the proper term for the concubine of a non-Native. She doesn't like being treated in any special way. She continues to stand firm on her actual situation—with a grandness rooted in revenge."

I still didn't intervene. Mama was being analyzed as if she were a character in a novel and Magda Peters was elucidating her personality in front of class.

"A person who is used to giving orders, running things, after giving everything proper consideration, Minke. She could run a much bigger business. I've never met a female entrepreneur like her. A degree from a Business Academy would not guarantee ability such as hers. You're right, Minke, she is a successful, self-educated person. And I've only talked about the business side. God! That's what's called a historical jump, Minke, for a Native. God, God! She should be living in the next century. God!"

I still only listened.

"In literary matters she could still learn from you, but even there she is, all the same, amazing. But you know what I found most amazing about her? She dared state her opinions! Even though there was no guarantee they were correct. She was not afraid of being wrong. Determined, with the courage to study from her own mistakes. God!"

I followed what she was saying without comment.

"I'd like to write about this extraordinary thing. It's a great pity I can't write like you, Minke. It's true what she said: without spirit, without fire. All I have is the desire, only the desire. Nothing more. You should be happy, Minke, being able to write. And that Association Theory, Minke, it lies collapsed and broken just because of that Native woman, your Mama, Minke. If there were just a thousand Natives like that in the Indies, Minke, these Netherlands Indies—Minke, these Netherlands Indies could just shut up shop. Perhaps I'm exaggerating, but it's only a first impression. Remember, first impressions, no matter how important, are not necessarily always correct."

She was quiet for a while then let out a long breath again. Her eyes no longer blinked nervously.

"She could advance even further. It's a pity that such a person would not be able to live among her own people. She is like a meteor shooting off by itself, traveling through space, knowing no bounds, who knows where eventually to come to ground, on another planet or back on our earth again, or to disappear in the infinity of nature."

"How much you praise her, miss."

"Because she's a Native, and a woman, and is indeed amazing."

"Please come again, miss."

"Pity. It's not possible."

233

"As my guest."

"It's not possible, Minke."

"Yes, Mama is always busy."

"It's not that. It seems your prima donna doesn't like me, Minke. I'm sorry. Thank you for the invitation. She loves you very much, Minke, that prima donna of yours. You should be happy, Minke. Now I know what all the gossip has been about."

14

I had come to feel at peace and safe at Wonokromo. Robert was never to be seen. Mama and Annelies never mentioned him. Even so it did not mean I could feel I'd taken his place. I put all my efforts into impressing people outside that I was not a bandit, and had no intention of acting like one. And that I was no more than a guest who could be told to leave at any time.

And one night after finishing some study I decided I would not do any writing. I would have a rest and then continue with some more schoolwork. I don't know why I had become so industrious all of a sudden. I wanted to get ahead at school. One thing was certain: It was not because of pressure from my family or from Annelies.

And I received no particular encouragement from my mother's letters either. They always asked if I was being beset by difficulties. I answered her fourth letter to tell her that I was doing well financially and to suggest that my monthly allowance be used for my younger brothers and sisters.

Correspondence was the hardest work. And in every letter I still gave Telinga's address. I only used the Wonokromo address when writing to Sarah and Miriam. They had started this. And I never asked them where they obtained the address.

I'd finished three algebra exercises that evening. The clock's pendulum struck nine times. The moment the chime stopped there was a knock on my door. Before I could answer, Annelies had entered.

"According to our rules you should be in bed by nine o'clock!" I rebuked her.

"No!" She frowned sullenly. "I don't want to sleep if Mas won't study in my room like before."

"You're becoming more and more spoiled Ann." Even Dr. Martinet would not have been able to handle a patient as difficult as she. I knew for certain: she really wouldn't go to sleep until her wishes had been one hundred percent fulfilled.

"Come on upstairs. Tell me a story until I fall asleep like you usually do."

"I've run out of stories."

"Don't make it impossible for me to sleep, Mas."

"Mama knows a lot of stories, Ann."

"Your stories are always better," and she closed all my books and pulled me up from my chair.

This doctor, obedient to his patient, gave in to her tugging, left the veranda, went upstairs, past Mama's room and the library, and once again entered Annelies's room. Over the last few days I hadn't been putting her blankets on or pulling down the mosquito net. As soon as she seemed to be getting well, she had to do all that herself.

She climbed straight up into her bed, lay down, and said: "Pull up my blanket, Mas."

"You're not really going to keep on being as spoiled as this?" I protested.

"Who else will spoil me if you don't? Now tell me a story. Don't just stand there. Sit here like you usually do."

So I sat on the edge of the mattress, not knowing what I must do near this newly recovered goddess of beauty.

"Come now, begin a beautiful story. One better than Stevenson's *Treasure Island* or *Kidnapped,* more beautiful than Dickens's *Our Mutual Friend*. Those stories don't speak, Mas."

I must always surrender for her health's sake!

"What kind of story, Ann? Javanese or European?"

"Whatever you like. I long for your voice, your words spoken close to my ear, so I can hear the sound of your breathing."

"What language? Javanese or Dutch?"

"Don't be so argumentative, Mas. Start a story now."

So I began to invent a story. I was unprepared. It couldn't just pop out of my head. And then I remembered the story of the love between Queen Susuhunan Amangkurat IV and Raden Sukra. It was a pity it was so frightening and wouldn't be good for her health. Dr. Martinet had left me orders: You must tell her happy stories that have nothing frightening in them. This child is amazing, he said further, even though her growth and intelligence are normal, her emotions are still those of a ten-year-old child. Be a good doctor, Minke. Only you can cure her. Strive so that she believes totally in you. She dreams of a beauty that does not exist in this world, perhaps because she's been forced to take on too many responsibilities too quickly. Her hopes are for a freedom without responsibility. Minke, such beauty, such incomparable beauty, must not die. Try. If you can become a vessel into which her trust can flow, only then will you be able to build up her self-confidence. Try hard.

So I began to make up a story as I went along. How it would end I didn't know myself. I'll plagiarize the characters at random, I thought. Let them complete their own tales.

"In a country, far, far away," I began,—"you're not being annoyed by mosquitoes, are you?"

"No. Why are there mosquitoes in that faraway country?" She laughed and her teeth glittered in the candlelight.

"In that far, faraway country there were no mosquitoes as there are here. Neither were there lizards crawling on the walls ready to eat them. Clean. That country was very, very clean."

As usual her gaze rested upon me. Her eyes shone dreamily as they did when she was ill.

"This country was fertile and always green. Everything that was planted thrived. Neither were there any pests. There was no illness, there was no poverty. Everyone lived happily and enjoyed life. Everyone was clever and liked to sing and dance. Everyone had their own horses: white, red, black, brown, yellow, blue, pink, green. Not one was blemished with spots."

"Kik-kik-kik." Annelies restrained her giggling. "There are blue and black horses!" she said to herself slowly.

"And in this country there lived a princess of incomparable beauty. Her skin was as smooth as ivory-white velvet. Her eyes were as brilliant as morning stars. No one could bear to gaze upon her for too long a time. A pair of eyebrows, just like slopes of mountains, protected that pair of daylight stars. The form of her body was what every man dreamed of. So all the country loved her. Her voice was gentle, overcoming the heart of all who heard her. If she smiled, the resolve of every man was shaken. And when she smiled, her white, shining teeth gave hope to every admirer. And when she was angry, her gaze focused and blood streamed to her face . . . amazing; she became more captivatingly beautiful still.

"One day she was going around the garden, on a white horse—"

"What was her name Mas, this princess?"

I hadn't yet found a fitting name, because I didn't yet know whether this story was taking place in Europe, the Indies, China, or Persia. So I continued:

"All the flowers bowed down, bending their stalks, shamed and defeated by her beauty. They paled, losing their splendor and their color. Only after the princess passed by did the flowers stand straight again, look up at the sun and complain, 'Oh, Great Sun God, why have we been shamed so? Did You not plant us on this earth as the most beautiful creatures in all of Your creation? You gave us the task of bringing beauty to humanity's life. Why is there now someone more beautiful than us?'

"The sun, too, was shamed because of these complaints and in his shame he hid behind a great, dense cloud. The wind blew, shaking the sad-hearted flowers. Then rain soon came and withered the leaves of all those multicolored flowers.

"The princess continued her ride, taking no notice of what had occurred behind her. The rain and wind could not find it in their hearts to disturb her. All along the road people stopped to admire . . ."

I saw that Annelies had closed her eyes. I took the mattress broom and shooed away the mosquitoes, then dropped down the mosquito net.

"Mas," she called, opening her eyes and holding on to my hand, preventing me from carrying out my intention.

I sat down again. The story had broken off. I groped around trying to start it up again:

"Yes, the princess rode on upon her horse. Everyone who observed her dreamed of how happy they would be if the gods would just turn them into the horse she was riding. But the princess did not know how they felt. She felt no different from anyone else. She never felt herself to be beautiful, let alone beautiful without peer."

"What was the name of the princess?"

"Yes?"

"Her name . . . her name . . ." she pressed. "Wasn't her name Annelies?"

"Yes, yes, Annelies was her name," and the story turned to tell about Annelies. "She had all kinds of clothes. Her favorite was a black velvet evening dress, which she liked to wear at all times of the day."

"Ah!"

"The princess longed for a beautiful love, more beautiful than had ever been glorified as having occurred among the gods and goddesses in the heavens. She longed for the arrival of a prince who was dashing, handsome, courageous, and more grand than the gods themselves.

"And then one day it happened. The prince she longed for arrived. He was indeed handsome. Manly also. But he did not have a horse of his own. Indeed he couldn't even ride a horse."

Annelies giggled again.

"He came in a rented buggy. There was no sword at his waist, because he had never gone to war. All be brought was a pencil, pen, and paper."

Annelies laughed again.

"Why are you laughing, Ann?"

"Was Minke the name of this prince?"

"Yes, his name was Minke."

Annelies closed her eyes. She still held my hand, afraid I would leave her.

"The prince entered the princess's palace as if he had just been victorious in battle. They chatted together. In no time at all the princess fell in love with him. There could have been no other outcome."

"No," protested Annelies, "the prince kissed her first."

"Yes, the prince almost forgot. He kissed the princess, and she went complaining to her mother. Not so her mother would be angry with the prince. But hoping her mother would approve of the prince's action. But her mother paid no attention."

"This time your story is wrong, Mas. Her mother not only paid attention. More than that. She was angry."

"Is it true she was angry? What did she say?"

"She said: Why complain? You yourself hoped for and expected to be kissed."

Now it was I who couldn't stop myself laughing. So as not to offend her, I hurriedly resumed the story again:

"How stupid was the prince. Twice now he has erred in his story. Indeed, the princess was hoping for and expecting his kiss."

"Liar! She neither hoped for nor expected it. She had no idea at all about what was going to happen. The prince came. He couldn't ride a horse, was even afraid of horses. He came, and before she knew what was happening, he kissed her."

"And the princess had no objections. Yes, she even forgot her sandals—"

"Liar! Ah, you're lying, Mas," and she pulled my arm hard, protesting against the untrue course of the story.

And so I fell into the softness of her embrace. My heart began to pound like the ocean whipped by the west wind. All my blood rushed to my head, dragging away my awareness and disrupting my work as a doctor. And I returned her embrace. And I heard her shallow breathing. And my own breathing too, or was it all my own, though I didn't realize it. The world, nature, dissolved into nothingness. There was only her and me, raped by a force that turned us both into a pair of prehistoric animals.

And we lay there exhausted, beside each other; we had lost something. All of nature suddenly went silent, without meaning. The pounding of my heart had stopped. Black clots began to arise within my heart. What was all this?

And Annelies held my hand again. Mute. And silently there was enmity between us. Enmity?

"Regrets, Mas?" she asked as I let go of a long breath.

And I did have regrets: an educated person with a mandate as a doctor. The clots of blackness were spreading everywhere. And indeed there were other regrets emerging that were not of my own will.

Annelies demanded an answer. She sat and rocked my body, repeating her question. I never guessed she was so strong. My answer was only to let out another long breath, longer still than before. She brought her face up close to mine to convince herself. I knew she needed an answer.

"Speak, Mas!" she demanded.

Without looking at her, I asked:

"Is it true I'm not the first man, Ann?"

She struggled free of me. Collapsed on the bed. Turned towards the wall with her back to me. She sobbed slowly. And I didn't regret having been so cruel, asking her such a tormenting question.

She was still sobbing and I still didn't react.

"You regret it, Mas, you regret it." She began to cry.

And I remembered my task.

"I'm sorry, Ann," and I stroked her thick hair in the manner she stroked the mane of her horse. She became quiet.

"I knew," she forced herself, "that one day a man I loved would ask me that." She became calmer and continued. "I have concentrated all my courage to be able to answer that question. To face it. I'm afraid. Afraid you'll leave me. Will you leave me, Mas?" Her back was turned towards me.

"No, Annelies darling," said the doctor.

"Will you marry me, Mas?"

"Yes."

She cried again. So slowly. Her shoulders shook. I waited until it receded. Still with her back to me, she spoke slowly, word by word, almost whispering:

"Poor Mas, not the first man. But it was not my wish, Mas, a disaster I could not avoid."

"Who was the first man?" I asked coldly.

She didn't answer for a long time.

"You'll seek revenge against him, Mas?"

"Who was he?"

"So shameful." Her back was still towards me.

Slowly but surely I realized: I was jealous.

"That animal." She pounded against the wall. "Robert!"

"Robert!" I answered viciously. "Suurhof? It's not possible!"

"Not Suurhof." Once again she pounded against the wall. "Not him. Mellema."

"Your brother?" I sat up, shocked.

She cried again. I pulled her roughly; she fell down on the bed. She covered her face with her arm. Her face was soaked with tears.

"Liar!" I accused, as if it was now my right to treat her in such a way.

She shook her head. Her face was still covered by her arm. I pulled her arm, and she pulled herself free, resisting me.

"Don't cover your face if you're not lying."

"I'm ashamed, Mas."

"How many times have you done it?"

"Once. Truly. A horrible accident."

"Liar."

"Kill me if I'm lying," she answered firmly. "Then one day you'll find out what happened. What's the use of living if you don't believe me."

"Who else beside Robert Mellema?"

"No one. You."

I let her go. I began to think over her shattering explanation. Was this perhaps the moral level of the nyai families? I almost answered yes. But I heard Jean Marais's voice again: Educated people must be just and fair, starting with how they think. I imagined Marais pointing and accusing: Your morals are no better, Minke. And I became ashamed of myself. She, Annelies, was no worse than Minke.

We remained silent for a long time. Each busy with our own hearts. Then I heard:

"Mas, let me tell you about it." Her voice was calm now. She needed to defend herself. Her sobbing had been replaced by a determined heart. Only she covered her eyes again with her arm.

"I still remember the day, month, and year. You can see it marked in red on the wall calendar. Almost half a year ago. Before I met you. Mama ordered me to find Darsam. People said he had gone down to the villages. I went to find him on my favorite horse. I went into village after village calling out his name. They— the village people—ran about to help me. He couldn't be found anywhere.

"Then someone told me he was inspecting the peanut crop in the fields. I turned and headed for the peanut fields. He wasn't there either. Even though there were no high trees, he was not

242

visible anywhere. His clothes, always black, made it easy for people to pick him out. And indeed, he wasn't there.

"A child I passed told me that he was on the other side of the swamp. Then I remembered: He was preparing a new field, one that was still dense with reeds. That field was to be planted with alfalfa and Job's tears for the new cattle Mama was importing from Australia. You couldn't see the field because it was surrounded by tall reeds.

"Do you remember that one remaining clump of reeds that I refused to go and look at with you?"

"Yes," and a picture of the bushes emerged; they were tall and bunched up together. She had refused. I could still remember how she had shuddered.

"I turned my horse in that direction while shouting out for Darsam from across the swamp. There was no answer. I came across a narrow track, broken here and there by bunches of reeds. And it was Robert I found.

" 'Ann,' Robert greeted me with a funny gaze. He'd thrown down his rifle and the string of birds from his morning's hunting. 'Darsam just passed here,' he said. 'He said he had to see Mama. He forgot he promised to see her at nine o'clock. He was two hours late.'

"I relaxed when I heard the explanation. 'And how many have you caught?' I asked. He fetched his string of birds and showed me. 'This is nothing, Ann,' he said again, 'the usual. I caught a strange animal today, Ann. Come down.'

"He walked a few meters and picked up the corpse of a big black-haired wildcat. I got down from my horse.

" 'This is not just any cat,' he said. 'Maybe this is the wildcat they call a blachan.'

"I patted the hair of the cat-victim, which had been struck on the head.

" 'No, I didn't shoot it. It was curled up asleep under a tree when I crept up and struck it dead.'

"His dirty hand grabbed my shoulder and I spoke angrily to him. He attacked me like a mad buffalo, Mas. I lost my balance and fell into the reeds. Had there been one sharp reed trunk, I would have been speared and would have died for certain. He fell on top of me. He held me with his left arm, at the same time covering my mouth. I knew I was going to be killed. And I struggled to be free, scratching his face. I couldn't fight those mighty muscles of his. I called out for Mama and Darsam. My calls died under the palm of his hand. Then I remembered Mama's warning: 'Don't go near your brother.' Now I understood, but it was too late. For a long time Mama had been alluding to his greed for Papa's estate.

"Then I realized he was going to rape me before killing me. He tore open my clothes. He kept my mouth covered. And my horse neighed loudly. How I begged that she would help me now. I closed my legs together like a vise but he prized them open with his powerful knees. I could not avoid the disaster.

"An unavoidable disaster, Mas," and she was silent for a long time again. I didn't say anything, but transmuted her story into images.

"My horse neighed again, came forward, and bit Robert on his bottom. My brother yelled out in pain, jumped up. The horse chased him for a moment. He ran out of the reed bush. I grabbed his rifle and ran out too. I shot at him. I don't know whether he was hit or not. In the distance I could see blood all over his pants, running down onto his trouser legs—the wound from the horse bite.

"I threw down the rifle. My body hurt all over. I tasted the salty taste of blood in my mouth. I couldn't climb up upon my horse, but when I neared the villagers I forced myself up on her so my disheveled clothes were hidden—"

"Annelies!" I exclaimed and I embraced her. "I believe you, Ann. I believe you."

"Your trust is my life, Mas. I've known that from the beginning."

Once again neither of us spoke for a while. It was then I had my doubts about the doctor's advice. She was adult enough. She knew how to defend herself, even if she hadn't been successful. She knew the meaning of death and trust.

"Didn't you say anything to Mama?"

"What good would that do? The situation would have got worse. If Mama found out, Robert would have been eliminated by Darsam, and then everyone would have been destroyed. Mama. Me. No one would come to our business anymore. Our house would have become a house of the Devil."

She spoke these last words strongly. But suddenly her strength disappeared: She hugged me again and cried again . . . and cried again. . . .

"Have I done wrong or not, Mas?"

I hugged her in return. And suddenly my heart pounded, whipped up by the east wind. And once again it happened; we formed a pair of prehistoric animals, until finally we rolled over on the bed again. This time there was no clot of blackness in my heart. And we lay in each other's embrace, like wooden dolls.

Annelies fell asleep.

Indistinctly, half-consciously, I thought I heard Mama come in, stop for a second in front of the bed, chase away the mosquitoes, and mumble:

"Hugging each other, like two crabs."

Half-awake, half-dreaming, I felt that woman pull up our blankets, pull down the mosquito net, blow out the candles, and then leave, closing the door.

15

My school friends still kept
away from me. The only one who began to befriend me again was
Jan Dapperste. All this time he was my admirer and looked upon
me as a *Mei-kind,* a "child of May," a child of good fortune, a child
who would never suffer failure.

He studied industriously, yet his marks were always below
mine. His pocket money also came from me. Perhaps because of
that pocket money he looked upon me as his elder brother. We
were in the same class.

Jan Dapperste always told me of any rumor about me. So I
knew all the ill-intentioned things Robert Suurhof planned against
me. From Jan I also found out that Suurhof had reported me to the
school director. Who cares, I thought. If they want to expel me,
let them. I can't do anything in this school anyway. Elsewhere?
Free and able.

Once the school director did call me in and ask why I had
become such a loner and didn't seem to be liked by the other
students. I answered that I liked them all, and that there was no
way I could make them like me. He then said that there must be

some reason they don't like me. Of course, Director. What's the reason? he asked again. I really don't know, I answered, I only know that there have been rumors about me spread by Robert Suurhof.

"Because you're not one of them any longer. Not a part of them, not the same as them."

I quickly understood: This was a sign I was to be expelled. Very well—I'd prepared myself to face such an eventuality. No need to be afraid. I may not continue my studies? No matter. In the final analysis school was no more than just a way to fill in time anyway. If I could advance, good; if not, no matter.

"We hope that you can improve your behavior. You will be important one day, an official. You're receiving a European education. You should be continuing your studies in Europe. Don't you want to become a bupati?"

"No."

"No?" He stared at me more sharply for a moment. "Ah yes, perhaps you want to become a writer. Or a journalist. Even so, proper behavior is still required. Or do I need to write a letter to the Bupati of B—— or Assistant Resident Herbert de la Croix?"

"If you feel that to write to them about me would be useful, of course there'd be no harm in doing so."

"So you agree that I should write the letters?"

"It's no concern of mine. That's your affair, Director. It has nothing to do with me."

"Nothing to do with you?" He gazed at me again, still more sharply. Then in amazement he resumed hesitantly. "So who am I talking to now? Minke or Max Tollenaar?"

"They're the same, sir, the one individual with different names."

He told me to go and didn't summon me again.

Miss Magda Peters also seemed to keep her distance, though she still seemed friendly, and I met her only during classes.

The school discussions were still suspended by the director.

And amazingly, I felt that whatever was going to happen, I was dependent on nobody. I felt strong. My writings were being read by more and more people. More and more were being published—even though they hadn't brought in a single cent all this time. If the public knew I was a Native, I thought, maybe their interest would dissolve away; perhaps they would also feel deceived. Only a Native! I was prepared to face this also. Jan Dapperste had already warned me of Suurhof's plan to expose me to the public.

The interview with the school director was not the only thing that occurred during that month, however. Not long after Jan Dapperste's whispered warnings, the paper *S.N. v/dD.* summoned me to their office. The managing editor of the paper wanted to meet me.

Jan Dapperste did not refuse my invitation to accompany me.

Mr. Maarten Nijman received us both and pushed across to me a reader's letter. Exactly as Jan had said: Max Tollenaar is only a Native. Jan and I recognized the writing. Jan nodded knowingly.

"Have you called me to obtain recompense because of this letter?" I asked.

"So what it says is true?"

"Yes."

"Well, indeed, we must make a claim upon you." He smiled agreeably. "We've prepared our claim. You no doubt know what we are going to demand?"

"No."

"Mr. Tollenaar, we demand that you work part-time for the paper, part-time but on a permanent basis." He pushed across a receipt and I received an honorarium for my past writings, even though it wasn't all that much. "From now on, as you're now a permanent helper, you'll receive more."

"How can I help?"

"Write, write whatever you like; and may you have success, sir."

The buggy took us to a restaurant. Jan Dapperste congratulated me and ate with great gusto, as if he had never eaten in all his life.

The third thing that occurred was a meeting with Dr. Martinet. It happened straight after we left the restaurant. Jan Dapperste was still with me. The doctor was waiting for me on his veranda, and said he wanted to see me.

"And, Doctor," he addressed me, "how is your patient?"

"Well, Doctor."

"What do you mean?"

"She's healthier now, working as before, and reads a lot in her spare time. She rides her horse when visiting the fields or the villages. She adheres strictly to the reading schedule I've prepared. Sometimes the three of us sit together listening to music from the phonograph."

"True. She seems well."

And Jan Dapperste was left alone on the veranda.

"Seems? So she's not as well as you hoped for, Doctor?"

"It's like this, Mr. Minke. I've already examined her five or six times over the last few weeks. I didn't notice at first. But after a third examination, I realized that she always shuddered and the hairs on her neck stood up whenever I touched her. Since then I have been suspicious. What is going on inside the body of this beautiful girl? I began to think there might be something amiss in her unconscious. I quickly started to study it. At first I thought she was revolted by me. Perhaps in her eyes I looked like an animal. Perhaps I was truly revolting. I examined myself in the mirror. I examined my face in great detail. No. I hadn't changed much over the last ten years except that I now wear a monocle in my right eye. Isn't my face normal and indeed, perhaps, handsome, even if only a little?"

"Not just a little, Doctor."

"Ah! A little is enough. You're the one who is really handsome. That's why she's chosen you and not me."

"Doctor!" I exclaimed, protesting.

"Yes, Dr. Minke," he laughed. "Only after I met you did I realize that she didn't shudder because of my appearance. It seems she shuddered because of my skin. White skin."

"Her father was also white-skinned. A Pure-Blood."

"Ts, ts, this is only a guess. Her father was white-skinned. Pure-Blooded. Yes. Listen, I called you to help me solve this problem. Yes, her father was a Pure-Blood European. Precisely. How many children in this world feel revulsion towards their parents? Deeply or not? Permanently or only sometimes? There are indeed no statistics, but there are such children, and not a few. This revulsion can be caused by the parents' own behavior, for example. If her parents had the same color skin as she has, she wouldn't be revolted by skin color."

"Annelies is also white."

"Yes, but with a Native softness. I myself have dreamed of her becoming mine. Funny, isn't it, Dr. Minke? It's a pity she's too young for me. Only a dream! Don't be angry. I'm not serious.

The fact is that she's revolted by me. Yes, she does have white skin. I'll make the following guess: An outside influence, strong and beyond opposition, has given her a false image of herself. She feels herself a Native, a genuine, real Native. She has obtained a picture from her mother that all Europeans are disgusting, loathsome, and act basely. My interviews with Nyai and Annelies encourage me to come to such conclusions. Nyai is indeed extraordinary. I think everybody acknowledges that. I've said before to you that she has unconsciously self-educated herself? And because of that she failed in another field? She doesn't understand how to bring up her children. She has placed them in the middle of her own personal conflicts. It's not just a deficiency—it's a failure, Mr. Minke."

It seemed that the conversation was going to go on for a long time. I excused myself for a moment and ordered the buggy driver take Jan Dapperste home.

"That child, who knows nothing, accepts everything that is crammed into her as part of her very self," he continued.

"But Mama doesn't hate Europe. She does a lot of business with Europeans, and with professionals like yourself. She even reads European literature."

"True. As long as it fits in with her interests. Just look at her relationship with Mr. Mellema. She progressed because of her master, but her unconscious always had its reservations and distrusted him. Everyone among elite circles knows the tragic story of Mr. Mellema and his concubine, except, perhaps, Annelies herself. Without being conscious of it, Nyai has molded Annelies into her second personality. That child will never show any initiative if far from her mother. Initiative, in the form of commands that Annelies cannot refuse, will always be something that comes from her mother. Have pity on that beautiful child. Her psyche is in confusion, Mr. Minke. Her mind is inside her mother's head."

I listened, staring, confused. It was an explanation that was involved, difficult, the first of its type that I'd ever heard, but clear and interesting. It was amazing how someone could peep into somebody's inside like peeping into the inside of a watch.

"The mother's personality is overwhelming, she knows so much: more than enough for her life's needs in this jungle of ignorance which is the Indies. People are afraid to face her, afraid that they will be able to move once under her influence. I, too, often find myself at my wits' end to know what to do. If she was only an ordinary nyai, then with that sort of wealth, with that sort of beauty, with an uncertain man, there would have certainly already been many thrushes coming around showing off their beautiful whistling. But no. None at all. None have come. None are singing—as far as I know, Pure, Indo, let alone any Natives, who most clearly of all would never dare try. They all know they would be facing a tigress. With one growl from her, such a company of crickets would disappear, tumbling head over heels in a daze."

"Is all this true, Doctor?"

"You must help me think it out."

"Is it proper for an H.B.S. student to be involved in work like this?"

"Ts, ts, ts, you're the one who has something at stake here. And, Dr. Minke, do you think I'm just making up a fairy tale? You're educated: Try to prove me wrong. That's why I needed you here. You are closer to them. It is you who must investigate what is happening; it is you who must understand what's happening. I'm just trying to give you a starting point. You are an adult. Furthermore, only you can be her doctor now. Not Martinet. She, that girl, loves you; and love comes from a source of power that has no equal. It can change people, destroy them or cause them to cease to exist, build them up or smash them down. It's her love for you that I put my hope in; my hope that she will be able to free herself from her mother, so she can develop a personality of her own. From my most recent observations, from her deliriums, from the look in her eyes, it is clear that she has surrendered her fate to you. This is no mere guess, not just some wild supposition."

His analysis was becoming increasingly interesting—because it did indeed relate to my personal concerns.

"Once she begins to say no to her mother, it means that there is some movement towards change occurring inside her. Of course one result will be pain, just as with all births. Nyai herself has unconsciously prepared the way for such a birth within the psyche of her daughter: She hasn't opposed Annelies's relationship with you, and has rather suggested it and made positive proposals, and even enthusiastically encouraged it. But there is still something else that is troubling her."

(There then followed, I think, a number of sentences which I couldn't quite understand because of my own limitations. I haven't noted these down here.)

"Annelies, your sweetheart, carries some burden that weighs heavily on that fragile heart of hers. All roads have been opened to you, cleared by her mother. It appears that it is you, Mr. Minke, whom Nyai wishes as her son-in-law. And it looks as if you approve of Nyai's wish. Even so, whatever it is that is burdening the girl's heart should also be of concern to us. She has been able to capture your heart, if I'm not mistaken. She should have the right to feel happy. But no, Mr. Minke. On the contrary, she is suffering very, very greatly: afraid of losing you, whom she loves with all her heart. This is piling up all sorts of sufferings on her. She could go mad, Mr. Minke, I'm not joking, a person could go insane, become totally unbalanced, lose her mind, go crazy."

He stopped talking. He took out a handkerchief from his pocket and wiped his face and neck.

"Hot," he said. Then he stood up and went over to the corner of the room where he wound up the spring of the fan. After it started up and began cooling down the room he came back and sat down. "For me, if treated just as an intellectual problem, this is all very absorbing; on the other hand, to see such youth and beauty dominated by uncertainties and fears is so saddening . . . do you understand what I mean?"

"Not yet, Doctor, these fears . . ."

"We'll come to them in a minute. Perhaps ever since Eve, beauty has excused the deficiencies and imperfections of people. Beauty lifts up a woman over her fellow women, higher, more honored. But beauty, and indeed even life itself, is all in vain if it is dominated by fear. If you still don't understand, I'll tell you what the problem is: She must be liberated from her fears, all her fears."

"Yes, Doctor."

"Don't just give me yes, yes, yes. You are an educated person, not a yes-man. If you're not of the same opinion as me, then speak up. It's by no means certain that what I'm saying is correct. I'm not a psychologist by training. So if you have different opinions then say so frankly so that our work of curing her will be easier."

"I have no opinion at all, Doctor."

252

"Impossible. Come on, out with it." I remained silent. "It's not too hot now, is it? Look, Mr. Minke, in science, the word *embarrassed* has no place. People should not be embarrassed of being in error or doing something incorrectly. Errors and mistakes like these will consolidate the truth and so also assist in our researches."

"It's true, Doctor; I have no opinion on this."

"I know you're trying to hide something. Educated people always have an opinion, even if a mistaken one. Come on, out with it."

His transparent, marblelike eyes gazed at me. He placed his two hands on my back.

"Look into my eyes, speak frankly. Don't make things more difficult for me."

I gazed at his eyes, and because of their transparency it was as if I could see through into his brain.

"With respect, tell me. Please don't make me fail in this work of mine."

"Doctor," I began, "truly, this is the first time I've ever heard an analysis of this kind. I'm still amazed by it all; how can I come to any conclusions? About Mama and Annelies, yes, I have often felt that there are some problems. Especially about Robert. My feelings about all this, my feeling mind you, not my opinion, or at least not yet my opinion, is that there isn't much wrong in what you've just said. On the contrary, I think it will help me to understand. Am I mistaken?"

"Good enough and not mistaken. In science, humility is sometimes needed. But only sometimes. But you don't need to be humble when answering my questions. But, yes, forgive me if it seems I'm acting like a prosecutor. I'm absolutely sure that all this is in your interests too."

The spring in the fan had almost wound down, so he went over to the corner and wound it up again.

"Good," he said, without sitting down. "Well now, please listen: What I'm about to say may be of some use to you in any deliberations at home. First of all, about her fear of losing you. This whole matter is in your hands. No one else can assist. As soon as she sees signs that you are going to leave her, she'll begin to become anxious. So you must not let her see any such signs, let alone actually leave her. To leave her would mean to break her."

He took a pencil from his desk. "Like this," and he snapped the pencil in two. "This broken pencil can still be used. But not a broken psyche, Mr. Minke. If such a person continued to live, she would be a burden upon all. If she dies it would be to everyone's regret. Haven't I already said you are her doctor? It could happen that, should you wound her love, you might end up killing her instead. Now I've told you, as clearly as I could. Without embarrassment, without fear, without self-interest. It's up to you now whether you become her doctor or her killer. By telling you this my own responsibility diminishes."

Now he sat down again. He put the broken pencil on the table. Then he gazed at me again, perhaps to convince me he wasn't just joking.

"Yes, Doctor."

"On the other hand, Mr. Minke, it is precisely because she has fallen in love with you that she is beginning to be born as a personality in her own right, because she is being confronted with a problem that is totally personal. This time no one else can give any commands. It is her birth as a personality in her own right that has made her fall ill."

I didn't understand any of this. I gazed into his eyes. I don't know why but all of a sudden I felt a surge of suspicion towards him, as a European. He seemed to know what I was feeling and hurriedly added:

"Once again, it is by no means certain that I am correct in what I'm saying, even half correct, let alone wholly. But while you have no opinion of your own, it would be wise for you to accept what I've said as a guide. So you do not become confused. A temporary guide."

He didn't resume his lecture for some time. I reckoned he was having doubts. The thought gave me great pleasure. At the very least I could breathe easily again. It was true, of course, that they were only words that he was pouring out at me. But it felt as if I were the anvil upon which he was hammering out some new understanding.

"Yes, Doctor." It was naturally I who began, but only as an indication that I was not an anvil without spirit.

"Yes." He spoke as if it were a complaint. And a heavy exhalation escaped from his chest, which was tight with problems. "Yes, this is all just conjecture at the moment, conjecture based on a number of facts," he resumed in defense of himself, asking for forgiveness at the same time. "I won't say any more until you have your turn: Now you must talk to me. In what room do you sleep?"

He knew I could not hide my embarrassment. Even at school such a question would be considered insolent; no one would ever question me like that.

"In science, embarrassment has no value, not even a tenth of a cent. Mr. Minke, help me. Only the two of us together can rid her of those fears. So, where do you sleep?" I didn't answer. "Very well. You are embarrassed—a feeling without value. But so my conjecture has been proved. So Nyai does desire the safety of her daughter. That is why you are embarrassed to tell me. You have already slept in the same room with her. I'm not mistaken?"

I could not look at his face anymore.

"Don't misunderstand me," he said hurriedly. "I have no desire to interfere in your affairs. For me the only important thing is Annelies's well-being, her well-being as a patient—and there-fore also, of course, your own and Nyai's well-being. All I hope from you is assistance. Aid in understanding. My guess needs to be verified. That is the only medicine for her. Your personal confidences and those of all my patients are safe and secure with me. I will always be a doctor, you are one only temporarily. Tell me about it."

To give me a chance to put myself in order, he went out to the back room for a minute. He returned carrying lemon drink and poured some into a glass for me.

"Why do you serve this yourself, Doctor?"

"There is no one else in this house. Only me."

"No washer or houseboy?"

"No."

"You do everything yourself?"

"A servant comes and works three hours a day, then goes."

"Your food?"

"A restaurant looks after that. Let's continue. Have a drink first. I know you need to get up some courage." He smiled sweetly.

And I had no courage.

"When the need arises," he began to advise me, "you must dare to learn and learn to dare to look upon yourself as a third

255

person. I don't mean a third person in the grammatical sense. Look: As a first person you think, plan, give orders. As a second person, you consider, you reject, or on the other hand you could approve, accept what the first person suggests. And the third person—who is that?—that is you as somebody else, as a problem"—he tapped the table with his fingertips—"as an actor, as somebody else that you see in the mirror. Tell me now about yourself as the third person, as seen in the mirror by yourself as first and second person."

"What must I tell you?" I asked once again.

"Anything at all about your relations with your patient."

"How must I begin?"

"So you are willing. Let me lead the way into the problem. Actually it's not that you cannot begin, but rather it's because you, as second person, are not fully willing. Let us begin. You have begun to live in the same room as Annelies. Now, continue on yourself."

"Yes, Doctor."

"Excellent. Nyai has never forbidden you or become angry because of this?"

"You are not mistaken, Doctor."

"It's not I who is not mistaken, but Nyai. She has done the better thing in looking after the well-being of her daughter. So she carried out that advice. Let's go on further. You sleep separately or in the same bed?"

"Not separated."

"When did this begin?"

"Two or three months ago."

"Long enough to get to know Annelies's major fears. Have you had intercourse with Annelies?"

I shook.

"Why are you shaking? Listen well: Here the problem is the important thing. Who knows if a similar problem will confront us some day in the future? Do you need another drink?"

"Excuse me, Doctor, I need to visit the toilet."

"Of course," and he showed me the way.

I met no one inside or outside the house. Still and silent like a graveyard.

Once in the bathroom I washed my face. I wet my hair. I felt the refreshing coolness of the water and my heart was refreshed too. I wiped away the dripping water with my handkerchief. Then I used the comb and mirror that was there. That is the third Minke.

As soon as I had sat down before him, he resumed:

"The more you try to hide things, the more tense you will become."

He was becoming increasingly able to peep into my psyche. I became nervous again. And there was nowhere I could hide my face.

"Come on. Give me reason to be even more thankful to you. I don't need to ask questions anymore, do I? You can continue the story by yourself."

I shook my head. I couldn't.

"Very well, if you still need a guide. You have slept in the same bed with her. You have had intercourse with her. Then you found out she was not a virgin. You had been preceded by some one else."

"Doctor!" I exclaimed. Without my realizing it, my nerves gave way and I began to sob uncontrollably.

"Yes, cry, Minke, cry like a baby; still innocent like at the time of your birth."

Why was I crying like this? In front of another person? Neither my mother nor my father? What was it that was worrying me? Perhaps I did not want my secret, our secret together, to be found out by others?

"So my guess was right. *Truly* you do love this girl. Her loss is your loss. You have lost something, and you want to hide that loss from the world. She was no longer a virgin. Yes, keep on crying, but answer my question. It's not the last question. It's important to get a picture of Annelies's first sexual relations in order to understand its influence upon her. A person's first sexual relations are soldered into the emotions of human beings, and can indeed determine a person's sexual makeup. No, no, that's not quite right. I should have said: It can determine a person's sexual makeup in the future. Now the question is: Has Annelies ever said or even been willing to say who it was? That first one? Or more accurately the first of those before you?"

"I can't, Doctor," I exclaimed in my pain.

"The third Minke must come forward. And it's not the last question yet. Who was he?"

I didn't answer.

"So you do know who he or they were?"

"Not them, Doctor, him."

"Very well, *him!*" He closed his eyes as if absorbing something. Then that question of his, which I did not wish at all to hear, struck like a thunderclap bringing me to consciousness: "Yes, him. Of course, him. Who was he?"

"Ah, Doctor, Doctor!"

"Very well, you don't need mention his name. Do you consider this person to be a good person or not? I don't mean in relation to sexual passions but in his day-to-day behavior."

"I don't dare, Doctor, I haven't the right to judge."

"It seems that you consider everything to be a personal secret or a family secret, or a secret of your future family at least. Yes, it's very moving—this loyalty of yours to all the members of your family or your family-to-be." He looked the other way as if to give me some freedom to use my own face. "At the very least I can make a guess at who the person was, especially now that I've seen your reaction. You are still young, very young, and it is you—even if only temporarily—who are Annelies's real doctor. So you must be strong. You are fond of her, even if you're unwilling to say you love her. I myself prefer to use the latter term. You have shown you are willing to accept the consequences of her deficiencies, to accept responsibility for her well-being. No matter what happens, you will not let go of her, because thousands of eagles will destroy her. Her beauty is indeed extraordinary, that Creole beauty which conquers people, no matter what country they come from. You will end up making her your wife, whichever way you look at it. Be a good doctor to her, now, tomorrow, and forever. The older we become the more complex becomes this life which confronts us, so the more courageous we must be in facing it."

The longer he spoke the clearer became the vision of Robert Mellema, scorning and insulting me, threatening me, glancing out of the corners of his eyes at me and waving his fists.

"Yes, it's your own reaction which has confirmed my guesses. If you're unprepared either to confirm or deny them, what can I do?"

"Doctor, Doctor . . . her own brother, Robert Mellema."

The glass of lemon drink in my host's hand fell to the floor, broken. I jumped up from my chair and ran out to my buggy.

Dr. Martinet visited us several more times. Usually he came in the late afternoon when Nyai Ontosoroh and Annelies had finished work. They all sat in the front yard chatting and listening to the music from the phonograph. I could see his buggy enter the compound and would come out to meet him after I had bathed.

After that earthshaking interview, about which I never told anyone except my diary, my respect for him became deeper and more unqualified. I didn't just look upon him as a doctor of great skill, a scholar of great humanity, but also as somebody who was able to plant the seeds of new strength within me. How he strived to understand other people! Not only to understand, but to hold out a helping hand—as a doctor, as a human being, as a teacher. He was a friend of humanity—a title that Magda Peters was to use for him some time later. He could show his friendship in so many different ways. And no matter in what way he did it, people were moved to put their trust in him. Sometimes I felt ashamed that I had been suspicious of him, even though it was my right to feel that way.

After observing him for a long time, my estimation of his age changed. Not in his forties, but in his fifties. His face was always that fresh, reddish color, and young. No lines of age disfigured his face. Every one of his statements was interesting and had content. He was a very clever storyteller, and without their knowing, he noted down people's reactions to his stories as a way of becoming acquainted with and understanding his patients. That's my guess anyway. I could be wrong.

During a visit to the house of an important citizen to arrange an order for a family portrait, I found that person reading an English magazine. When he went inside to fetch something, the magazine was left lying open on the table. Indeed a coincidence. And even more a coincidence that I needed to take a peep into the magazine and found an article by Dr. Martinet inside. Its title was "The Beginning of a New Age of Social Transformations as a Source of New Illnesses." In a box there was a paragraph that announced that any therapy that did not take into account the patient's social background was still primitive.

The man returned and I put the magazine down again. Since that moment I knew that Dr. Martinet was also a writer. Not a writer of stories like me, but a scientific writer.

And when he arrived that afternoon I tried to observe his behavior more carefully. I no longer needed to fear him, scared that he would peep into my psyche.

As usual his story that time also contained a message, even though it was told jokingly. It was about twins who, ever since they were children, ate from the same plate and drank from the same bowl. As soon as they entered adulthood, though their faces were identical, they became different people. Each was moved by different desires and dreams. Their desires and dreams shared the same origins. They were born out of an unfulfilling reality and also of differing images of themselves, of what they wanted to become.

At first I didn't understand what he was getting at. Mamma and Annelies didn't say anything either. Perhaps they were bored, but then he added:

"Like Miss Annelies here. She has everything: money, a mother who loves her, incomparable beauty, many skills. But there is still something that she feels she does not or does not yet have. You must recognize that she has such a desire. If not it will become an illness. And unconscious desire can govern one's body with great viciousness, showing no pity. Both emotions and thoughts are controlled by it, governed by it. If such a desire is not recognized people can behave as if ill—as if disturbed. Miss, what is it that you want so much that you have fallen ill?"

"There isn't anything. Truly there's nothing."

"And so why have you all of a sudden gone red? Isn't it true you want Mr. Minke?"

Annelies glanced at me, then bowed down her head.

"Nyai, if I may make a suggestion, marry these two at the very first opportunity." He looked straight at me. "And Mr. Minke, you have learned to dare? Learned to be strong as well as daring to learn?"

He didn't continue. A rented carriage entered the compound. The driver helped his passenger down: Jean Marais. May jumped down, then led her father along.

I introduced them to the others:

"Jean Marais, artist, designer of household furniture, French nationality, my friend doesn't speak Dutch."

The atmosphere changed. The problem was that Dr. Martinet didn't understand Malay. Mama and Annelies didn't understand French, even though Dr. Martinet did. Only May and I knew all their languages. And May quickly stuck to Annelies.

Dr. Martinet nodded his head seeing Annelies's joy in gaining a younger sister while May obtained an elder sister. In a second he directed his eyes at Jean, and asked in French:

"How many children do you have?"

"There have been no opportunities for May to have brothers or sisters, Doctor," and his answer and his eyes radiated his displeasure at having to answer that question.

But Martinet, with his practice of piercing through into another's psyche, paid no attention and continued in Dutch addressed to no one in particular:

"How beautiful it would be if it were possible for the two girls to get together. It should have happened long ago."

Meanwhile Annelies had taken May inside the house. They didn't come out again. Their laughter and chatter could be heard in the distance, sometimes in Malay, sometimes in Dutch and Javanese. Jean Marais shook his head as he listened to his child's voice. His face shone. But the atmosphere still remained awkward. Dr. Martinet felt uncomfortable. He excused himself, and climbed aboard the carriage that was waiting for him beside the house.

"Mr. Marinet is a very clever Doctor," I said in Malay. "It was he who cured Annelies. We are very grateful. This friend of mine here, Jean Marais, has come to ask permission to paint Mama, if Mama agrees and has time."

"What's the use of being painted?"

"Madam," Jan responded.

"Nyai, sir, not madam."

"Minke greatly admires madam—"

"Nyai, sir."

"—as an extraordinary Native woman. He is always singing madam's praises—"

"Nyai, sir."

"—so we are in agreement that they should be made eternal in a picture. In the future, who knows if in one or forty years, people will still know who you are and be able to admire you."

"I'm sorry, but I have no desire to be admired."

"That can be understood. Only stupid people admire themselves. But it is not madam herself who admires madam, no—but rather the living witnesses of the age."

"It's a pity, sir—I'm not willing. Not even to have my photo taken."

"If so, yes, it's indeed a great pity. If so—if so—may I then just look upon madam so as to memorize you in my heart?" he asked politely and awkwardly. Nyai went red. "So that I can do the painting at home."

Nyai's gaze swept from me across to the house, then to the signboard in the distance. Finally it settled on the garden table. She looked disturbed, embarrassed, and her movements were awkward.

"No, no sir." She was embarrassed. "And you, Minke, what have you been saying about me outside?"

"Nothing bad, Madam. Only praise."

Seeing the confusion Nyai was in, I spoke up quickly: "Mama doesn't want to be painted now. Perhaps at some other time."

"Not at any time."

"He's my friend, Mama."

"Then he's my friend too."

Jean Marais, who had always been sensitive, perhaps because of his deformity, looked nervous and as if he wanted to leave quickly. His eyes nervously sought out his daughter, but only her voice could be heard singing in the distance.

"She is inside, sir," said Nyai. "Please come in."

We entered. The gay singing of May and Annelies became clearer. And Nyai seemed to be very happy to hear it. I'd never heard Annelies singing while I'd been at Wonokromo. It seemed as if she was returning to her childhood, a period that was far too short, torn away from her by responsibilities and work.

Jean was lost in silent daydreaming.

"Mr. Marais," said Mama after we had all been sitting silently in the front room. "Your child, it seems, has brought a gust of fresh air to this house. What about if she comes here often, just as Dr. Martinet suggested?"

"If the child wants to, I can see no reason why she shouldn't." His voice was despondent, as if he was afraid of losing something.

"Minke, Nyo, invite Mr. Marais to stay overnight."

"What about it, Jean, would you like that?"

For the umpteenth time I saw how awkward this artist was, this creator of beauty. He couldn't even answer such a simple thing. He gazed at me, not knowing what to do.

"Yes, Jean, you should stay the night. Tomorrow, early in the morning, I'll take you back to the workshop so you don't have to open late."

He nodded in agreement, forgetting to say thank you for such a friendly invitation.

That night as we lay in the same bed before going to sleep, trying out Dr. Martinet's method of conversing, I asked:

"Jean, you seem dispirited lately. Are you still lamenting over your past? I'm sorry."

"That is the question of a writer, Minke. You're truly a writer now, one hundred percent."

"It's not that Jean. I'm sorry. I'm far, far younger than you, Jean, with far less experience and knowledge. Will you answer, Jean?"

"It's a very personal thing. And moreover I'm going to close the matter with the completion of that painting I started a while ago. Are you going to write about me?"

"You are a very individual person, Jean. Yes, providing I can do it properly. What is it that you really desire, Jean?"

"Desire? Ah, you! You are an artist. I am an artist. Every artist desires—dreams about—reaching the peak of his success. Success! And he gathers together all his energy, Minke, only to defend that success, a success that always torments and oppresses him."

"But your voice is so despondent, as if you no longer believe that such success will ever come."

"Such a question! You are truly already an artist. I hope that question was born from your own spiritual struggles, a fruit of your own recent work. It is truly not the question of some-body your age. A question that contains authority. Do you be-lieve in your own questions?"

I stopped. I asked as confidently but casually as possible:

"What do you mean by authority?"

"In short: someone who truly understands his own ques-tions."

It was clear he wasn't sleepy. And it was clear too that my efforts to get him to open up had failed.

263

And on that night I submerged myself into so many problems. I felt I was saying good-bye to my youth, which had been
so gloriously beautiful and full of victories. Yes, even though they
mightn't mean anything to others. It was all these things that I had
noted down that gave me the right to claim victories. And the
greatest of these victories was Annelies's love. Even though, yes,
even though she was no more than a fragile doll.

The evening silence was broken only by the sound of the
pendulum clock.

Then I remembered a sentence once uttered by Dr. Martinet:

"Nyai's dairy cattle, in the process of becoming dairy cows,
fully developed cows, adult cows, need only thirteen or fourteen
months. Months! Human beings need tens, even scores, of years
before they grow into full adults, human beings at the peak of
their abilities and their worth. There are indeed those who never
grow into adults, who live off the handouts of other people and of
society: the insane and criminals. The resilience and strength—or
otherwise—of a person's abilities, and his worth, are directly related to the size and number of the trials he has undergone. Those
who always run from tests and trials—the insane and the
criminal—never reach adulthood. A cow is fully developed in only
thirteen or fourteen months—and without undergoing any trials,
any tests."

Ya Allah, in truth, the trials and tests You have made me
undergo have been too great for someone as young as me. My
situation has forced me to grapple with questions that should not
yet be my concern. Give me the strength to face every trial and test
You confront me with, just as You have done with others before
me . . . I am not insane. And neither am I a criminal. And never
will be.

16

That morning the sky was overcast. It had been a fine, clear week. It was only my heart that was not clear. The gray clouds that suddenly appeared within my breast warned me that a storm was on the way. Yesterday when I was out riding (all of a sudden I could ride quite well!) with Annelies—a Saturday afternoon with no school discussion—I glimpsed Fatso for a moment. Since then I had begun to feel anxious again.

I saw him on a cheap horse riding out of one of the villages on company land. In the evening, when Darsam came to my room to study reading and arithmetic, I refused to teach him. I told him there was a fat man who was acting suspiciously, who had followed me from B——. (Yes, I had suddenly remembered: He had bought a ticket at the railway station at B—— immediately after I bought mine. And I remembered too that he had arrived earlier and had hung around the platform talking to somebody.)

"Is he slant-eyed, Young Master?" Darsam asked.

"A bit," I affirmed.

"Yes, he's been seen several times now in the village," Darsam continued, and he thought he was an ordinary peddler.

"If he were a peddler, he'd have a pigtail for sure. He didn't," I said. "Maybe he's on orders from Robert."

Darsam didn't answer.

"Where is Robert now? He hasn't been seen since I returned from B——."

"He wouldn't dare come home. Do you remember what I told you before, Young Master? He ordered me to kill Young Master? And I said to him: My employers are Nyai and Noni; their friends are my friends. If Sinyo wants Young Master dead, it's best that it is Sinyo himself that I cut down. You're not my employer—look out! I pulled out my machete, and he ran."

It was yesterday that our conversation took place. The appearance of Fatso cast a shadow over my soul. And the morning sun was not able to cast out the gray clouds inside me.

"So you've seen Fatso?" I had asked Darsam the night before. "If you meet him again what will you do?"

"If it's true he's Robert's man, he'll feel the steel of my machete."

"Hush! Don't be crazy," I forbade. "You mustn't do that. If you did, everyone would get into big trouble. You mustn't, Darsam, you mustn't, understand?!?"

"I mustn't, Young Master, all right, I mustn't. But I'll beat him until all his bones are broken, so he won't be able to do anything again for the rest of his life."

"No. We don't yet know what the situation really is. If the police become involved, who'll help Mama? I can't. I'm not able."

And Darsam was silent. Then he spoke slowly and hesitantly: "All right, I will listen to Young Master."

"Yes," I said, "you must listen. I don't want to be the cause of some disaster befalling this family. And . . . we must ensure that no one else knows about Fatso."

And that morning I saw Darsam walking hither and thither restlessly. He was deliberately making his presence known so that I could call him at any time if I needed him. I knew: He was guarding my life from Fatso.

The three of us—Mama, Annelies, and I—were sitting on the front veranda listening to a recording of a popular song. The music jumped about like a school of river prawns at flood time. My heart was still enveloped by those gray clouds. I had a premonition: Something was going to happen.

I observed Mama and Annelies one after the other. And Mama was clearly suspicious of Darsam because of his unusual behavior.

"You seem uneasy, Mama," I said.

"It's always the same. If Darsam is running about like a kitchen mouse, I always become uneasy. Something always happens; I've been restless since last night. Darsam!"

And Darsam came and stood at attention.

"Why are you running about like that?" asked Mama in Madurese.

"These itchy feet of mine just don't seem to want to stay still; they keep moving about of their own accord, Nyai."

"Why aren't your itchy feet itchy out at the back?"

"What can I do, Nyai, these feet of mine keep taking me to the front."

"All right. But your face is so frightening. Harsh. Your eyes are wide open and are thirsting for blood."

Darsam forced himself to laugh exuberantly and left after raising his hand in respect. His mustache still waved up and down as if he were pronouncing some mantra. His eyes were indeed wide open today, as if his ears were capturing some mysterious voice from the heavens.

"Why are you so quiet, Ann?" I asked.

"It's nothing," and she rose and went over to the phonograph and turned it off.

"Why did you turn it off?" Mama asked.

"I don't know, Mama, the music just sounds like a lot of noise today."

"Perhaps Minke still wants to listen to it?"

"It's all right, Mama. Ann, do you still remember the man who was riding the horse yesterday?"

"Wearing the brown-striped pajamas?" I nodded. "Who is he?"

"Who was riding a horse? Where?" Mama asked hurriedly.

"In the village, Mama," Annelies explained.

"No one has ever visited the villages on horseback. Except for Mrs. Karyo's son, the watchman at D.P.M."

"It wasn't him, Mama. And he never wears pajamas when he comes home to visit his parents. This man was fat, clear *langsat*-colored skin, a bit slant-eyed."

"Darsam!" Mama called.

"Ah, see Nyai, that's why I need to have itchy feet!"

And Mama didn't respond to his jest.

"Who was the fat man on the horse in the village yesterday?"

"Just a peddler, Nyai."

"Nonsense. Since when do peddlers ride horses? Your behavior is strange today too. Even if he could rent one, he wouldn't know how to ride. Did he have a pigtail?"

Darsam, very unusually, laughed boisterously for a second time, trying to hide something. Then:

"Since when has Nyai lost faith in Darsam?" He wiped his mustache with the back of his arm.

"Darsam! You're really strange today."

And the Madurese fighter laughed again, saluted, and left without another word.

"He's hiding something!" Mama mumbled. "I feel more and more uneasy. Let's go inside."

Unable anymore to read, she stood up and went inside.

"Darsam, and Mama too, Mas, they're behaving so strangely," said Annelies. "Why?"

"How do I know? Let's go inside."

Annelies went in. I stood there surveying the scene, and then I saw Darsam running towards the main gate with his unsheathed machete in his right hand. And outside, for just a moment, I glimpsed Fatso walking along the road in the direction of Surabaya. He was wearing an ivory-yellow suit, white hat, and white shoes, and was carrying a cane, like someone out on a picnic. My earlier suspicion, that he could be a Majoor der Chineezen, no longer held.

On seeing Darsam I called out straight away:

"No, Darsam! No!" and I ran after him.

And Darsam didn't listen to me. He kept on after Fatso. I had no choice, I ran after Darsam to try to stop him. Nothing must happen. And Darsam kept on after Fatso. And I kept on running too, yelling out for him to stop—running with all my strength.

From behind me, I heard Annelies cry out:

"Mas! Mas!"

I glanced back for a moment. Annelies was running after me.

It seemed that Fatso knew he was being chased. He ran with all his might to save that abundance of flesh from the fighter's machete. Now and then he glanced back.

"Tso! Fatso! Stop!" Darsam shouted hoarsely.

Fatso bent down so he could run faster.

"Darsam! Come back! Don't go on!" I shouted.

"Mas, Mas, don't follow them!" exclaimed Annelies from behind me, shrilly and loudly.

I reached the main gate. Fatso was out ahead, heading straight for Surabaya. Darsam was getting closer.

"Anneliesss! Aaaaan! Anneliessssss! Come Baaaack!" Nyai could be heard calling out.

When I looked back I saw Mama, holding her kain up high, chasing her daughter. Her hair had fallen free and loose. Fatso was running to save himself. Darsam chased after Fatso. I chased after Darsam. Annelies chased me. And Nyai chased her daughter.

"Darsam! Listen to me! Don't!"

And he took no notice. He ran and ran. In a moment he would catch up with Fatso, who would then lose his head. No! It must not happen.

"Mas! Mas! Don't join in!" exclaimed Annelies.

"Ann, Anneliessss, come home!" exclaimed Mama.

And if Fatso had run on in the direction of Surabaya, he would have died for sure. The road was quiet on a Sunday, with just paddy, paddy everywhere, Ah Tjong's brothel, and Nyai's paddy, paddy and fields, and more paddy, and only then forest. It seemed he knew the area. His only chance: to turn into Ah Tjong's yard. He did it, and disappeared from my view.

"Don't turn!" ordered Darsam to his intended victim.

"Darsaaam! Alaaa! Darsam!" I exclaimed.

Then the fighter turned also and disappeared.

"Don't go in there!" came Nyai's indistinct shout.

"Don't go in there!" Annelies passed it on.

And now I too turned into Ah Tjong's compound. Fatso wasn't to be seen anywhere. There was only Darsam standing, confused, not knowing what to do next.

The front doors and windows were closed as usual. When I caught up with Darsam he was still panting. I too was out of breath.

"The rat has disappeared, I don't know where he's gone, Young Master."

"All right, let's go home. Don't keep on."

"No, he has to be taught a lesson."

There was no stopping him. He walked past the row of windows along the side of the house.

"Mas! Don't go into that house!" called out Annelies from her neighbor's gate. "Mama forbids it." But she herself had entered, tottering, into the front of the yard.

Darsam looked left and right. I pulled at him to make him return. He ignored me. His naked machete remained outside its sheath. In the end, I too became wild-eyed.

It turned out that Babah Ah Tjong's building was much bigger than it appeared to be from the outside. There was a long annex at the back. Almost all the surrounding grounds were garden, with fruit trees and flowers. They were all very well looked after. Everywhere could be seen thick heavy-looking black-painted benches. A narrow path, covered with layers of river gravel, cut up the yard into little sections.

For just a moment I caught sight of a couple. They didn't see us. Such views were never visible from outside, closed off by high, thick, multirowed walls.

Darsam turned right, circling the main building. There didn't seem to be anybody around. A back door was standing wide open. Behind me, Annelies had passed the row of side windows. Nyai's shouts could be heard more clearly:

"No, don't go inside!!"

And without hesitating Darsam went inside. He stopped, looked left and right, with his machete still in his hand.

And I too entered.

A large room, a dining room, opened up before us, complete with furniture: table and chairs, a buffet with all sorts of crockery inside. Mirrors painted with Chinese calligraphy hung on the walls. A few Japanese paper paintings of ocean prawns, bamboo, and horses also hung on the walls.

Suddenly Darsam was startled, and stopped dead in his tracks. His two arms shot out and stopped me from going any farther. I kept going on. What was there?

The body of a European lay in the corner of the dining room. The body was long and big, fat, large-stomached. Its blond hair was already threaded with gray and he was somewhat bald. His right hand was raised up on his head. His left hand lay on his chest. His throat and neck were covered in yellow vomit. The smell of liquor filled the room. His shirt and pants were filthy, as if they hadn't been washed for a month.

"Tuan!" whispered Darsam. "Tuan Mellema!"

Hearing that name I shuddered, and shuddered again as I approached the person with that familiar body, fatter than I had seen before, sprawled in the corner like a meditating ascetic. He was possibly in an extraordinary state of drunkenness or had fallen asleep after vomiting.

Darsam approached, crouched, and felt and pushed the body with his left hand. In his right hand his unsheathed machete was alert. The body did not move. Darsam then shook it back and forward, then felt the man's breast.

I came up close. It was indeed Mellema.

"Dead!" hissed the fighter. Only then did he glance at me, and continue his hissed speech: "Dead. Tuan Mellema is dead." And the frightening look on his face disappeared at once.

Annelies appeared at the door, calling hoarsely, out of voice, panting.

"Mas, don't go inside this house!"

I went outside, down the stairs, pulling her by the shouder. Mama arrived, also gasping. Her face was red and her hair was disheveled and all over the place, falling in a mess across her ears, face, neck, and back. She was soaked in sweat.

"Come on, come home! Everyone! Don't go into that accursed house!" she whispered, gasping.

"Young Master!" called Darsam from inside.

"Don't enter!" I now forbade Annelies and Mama. And I entered.

Darsam was rocking Mellema's body. The machete was still in his right hand.

"He's dead all right," he said, "he's not breathing. The blood has stopped too."

Annelies and Mama were suddenly behind me.

"Papa?" whispered Annelies.

"Yes, Ann, your Papa."

"Tuan?" whispered Nyai.

"Dead, Nyai, Noni: Tuan Mellema is dead," said Darsam.

The two women stepped closer, then stood still in a daze.
"That smell of liquor!" whispered Nyai.

"Mama?"

"Ann, take note of that smell," whispered Nyai again, without stepping closer. "Do you remember it?"

"Like Robert that time, Mama?"

"Yes, when he began to go mad too," continued Nyai, "and like Tuan the first time. Tuan went that way too. Don't get close, Ann, don't.'"

All of a sudden everyone looked up when they heard a woman's footsteps. And they saw a female in a yellow kimono patterned with big red and black flowers. Her skin was more white than yellow: a Japanese woman. Her quick, short steps brought her in our direction. Then she spoke to us in Japanese with a clear and attractive voice. We couldn't understand.

As an answer I pointed to the corpse strewn in the corner of the dining room. She shook her head and shuddered, turned right, and ran off with those short steps, more quickly, and went into the inner section of the house through a corridor.

We followed her with our amazed looks. That was the first time I had seen a Japanese woman. The round face, slanted, narrow eyes, the cherry-red, parted lips, one gold tooth: I don't think I'll ever forget it.

Not long after, out of the same corridor, there emerged the body of a tall man, an Indo, thin, with sunken eyes.

"Mama," whispered Annelies, "Robert, Mama."

Only then did I recognize that handsome youth, who had changed so much. It was indeed Robert.

Hearing Robert's name spoken, Darsam jumped up, forgetting Mellema's corpse.

"Nyo!" he shouted.

Robert stopped that moment. His eyes shot wide open. As soon as he recognized Darsam and saw the machete, he turned and ran. Darsam chased him.

Annelies, Nyai, and I were nailed to the floor. Dazed. For a second I imagined Robert sprawled out covered in blood, with a gaping stab wound. But no. Darsam came back again. He wiped his mustache. His face was wild.

"He ran, Nyai. Went into a room, jumped out the window. I don't know where to."

272

"Enough, Darsam, enough." Only then could Nyai talk. "Don't keep on with this craziness. He's my son." Her voice vibrated. "Look after your Tuan."

"Very well, Nyai."

Annelies held her mother's sleeve; she was shivering.

"See," Nyai hissed, holding back her anger, "nothing goes right. You go home, Ann. What did I say? Don't come into this house of sin. Pick up and carry back your Tuan, Darsam."

"Borrow a cart," I instructed Darsam.

Only then did the fighter sheath his machete and go outside.

Now Nyai stiffened as she looked at her master's corpse, while Annelies buried her face in her mother's breasts.

"Didn't want to be looked after properly. Preferred to be looked after by a neighbor. Ah Tjong! Ah Tjong!" Nyai called out. "Ah Tjong! Babah!" and the person being called did not appear.

Darsam entered again, frowning:

"The impudent caretaker won't lend us a carriage."

"Where's Babah?"

"He's not here, he said."

"Fetch our own carriage."

"Let me go," I said.

"You two stay here," said Nyai. "I'll go back. Come on, we'll go home, Ann!" and she pulled her child along.

The two women held hands, leading each other out of Mr. Tjong's pleasure-house through the back door. They took no notice of Mellema's gaping-mouthed corpse, sprawled out on the floor.

I saw then just how totally Nyai had broken with her master. She was not even prepared to touch him, even though he was the father of her children. She could never forgive him.

"Such a good beginning, such a hateful end, Young Master," Darsam grumbled. "What he hunted he lost, what he caught was cursed."

Soon after, there was the sound of uproar from the rooms. Women could be heard running about.

"Babah Ah Tjong's whores," hissed Darsam. "Five years Tuan nested here, here too he died. Dying in a whore's nest. Tuan! Tuan Mellema! Five years Nyai maintained her wrath. Even on his death, she showed no concern . . . human trash!"

273

"And Robert was here too."

"Under the same roof, with the same whores. Damned ones!"

"Mama had to pay for it all?"

"A bill came every month."

"Don't move the corpse," I ordered him.

A carriage arrived. Not Annelies, not Mama. Four police officers and their commandant, an Indo. They made an examination. One took notes of everything said by his commandant.

"Has he been moved?" the commandant asked in Malay.

"A little. I shook him," answered Darsam in Madurese.

"Where's the owner of the house?"

"Not here."

"Who lives here?" He took out his pocket watch, looked at it for a moment, and then put it back.

Not one of the house's inhabitants appeared.

"Who saw the body first?"

Darsam coughed, as his answer.

"What's the explanation of why the whole of the Boerderij household turned up here?" he asked in Madurese.

My heart pounded fast. There was no way of stopping it from becoming a police affair now. And all will be involved in difficulties.

"I was chasing Fatso."

"Who is this Fatso?"

"A suspicious character. He ran, I chased him, and he disappeared into here," Darsam explained.

"You entered someone else's house? Without permission?"

"There was no one here when we arrived. Anyone can enter here without permission. It's a pleasure-house."

"But you didn't come here for that."

"I've already told you"—Darsam was offended—"we came after Fatso. Perhaps a customer here."

The commandant laughed insultingly. And the other policemen lifted up the corpse. Not strong enough. Darsam helped, just to avoid more questions.

"Very well. What are your names?"

So Darsam and I were taken away in the government carriage. We were questioned more thoroughly at the station. And . . . in the end Father would indeed read his son's name in the paper—the cleverest among his children, the one the whole family was proud of, involved in a police case, and a dirty one—in a pleasure-house too—all just as Father had predicted.

That day we found out that Mr. Mellema died of poisoning. His vomit and the phlegm in his mouth and throat pointed to this fact. According to the investigations of Dr. Martinet, who was asked to conduct the autopsy, the poison had been given in low dosages over a long period, so that the victim had become used to it. On the day of his death he had received a dosage two or three times greater than usual.

And in the end it did happen. Reports began to appear in the daily press: the death of one of Surabaya's richest men, the owner of Boerderij Buitenzorg, Tuan Mellema, dead in Babah Ah Tjong's Wonokromo pleasure house; dying in poisoned alcoholic vomit! And our names were mentioned over and over again.

Reporters kept coming to our house: Native, Chinese, Indo, and Pure European. Mama and Annelies refused to answer any questions. It was I who forbade them to open their mouths. And in the street outside, people collected to watch us. Yes, we were beginning to be regarded as freaks on show.

None of us was detained. I used the opportunity to write a report of what really happened and it was published by the S.N. v/d D. A long time later I found out that my reports had increased the paper's circulation. People in other towns also sought that Surabaya paper because it was considered a credible source. The unnatural death of a wealthy man always gives rise to many suspicions and rumors.

I used my week's leave from school to write, to repudiate all the false and tendentious reports. But then there appeared another report, allegedly from police sources, that the police were carrying out investigations and were hunting for Fatso and Robert Mellema, who was the eldest of the Mellema children, both of whom were strongly suspected of conspiring to kill Robert's own father.

Who was this Fatso? One Malay-Chinese daily asked. The article mentioned the possibility that he could be a recently arrived Chinese illegal immigrant. Perhaps he was a member of that group calling itself the Chinese Young Generation, who wanted to overthrow the Empire. One of their special features: They wore no pigtails! And, indeed, Fatso wore no pigtail. Maybe he had come to Java because he was being pursued by the British police in Hong

Kong or Singapore. Now he was making trouble in Surabaya. Firm action needs to be taken against illegal immigrants, especially those without pigtails, who obviously had criminal intentions.

This guess was based on no more than a sucking of one's thumb! I replied to that Malay-Chinese paper. He indeed was slant-eyed—but that is not a characteristic unique to Chinese. He had no pigtail—but that too need not be interpreted as a sign he was a member of the Chinese Young Generation.

The result of my article was that the police questioned *S.N.v/d D.* about Fatso. Maarten Nijman refused on principle to give any explanation. Actually he didn't really know what it was all about anyway. For his refusal, he was detained for three days and nights.

Miriam and Sarah de la Croix expressed their sympathy to me, to us, and were sure that we were innocent of any wrong-doing. Greetings also came from Herbert de la Croix, who hoped that we would be able to face all our trials with strength and patience and that we would get through them all safely.

Mother's letters, so moving, told of her sadness, as well as telling me of Father's fury, which had reached such a peak that he actually said he no longer acknowledged me as his son. He had even written a letter to the director of my school withdrawing me from school!

In the next letter from Mother, also written in Javanese language and script, she said that it was not certain that I was in the wrong, and she hoped I would be the one who cleared the matter up. And that Assistant Resident B—— had visited father to calm him down and to pass on to him the above words; and also to say that my living at Boerderij Buitenzorg did not necessarily have any connection with anything indecent; that such a matter can occur as a result of one's own actions, but also can occur as a completely unrelated accident; no one can guess when such an accident, such a disaster, would befall them. Father did not contradict him. But to his sons and daughters he said: To become involved with the police is to shame and humiliate me, and whoever does become mixed up in a police affair is unfit to be near me.

I replied to all the letters. Responding to Father's pronouncement I wrote: If that is what is desired by Father, so be it; so from now on I will devote myself only to my mother.

My elder brother wrote to me: Mother was bathed in her own

tears on reading your reply. She cried over your attitude, why did you approach your father, who was already so furious with you, with such lack of devotion, as if he was a father who had never wished anything good for you, his own son. You are his son, you are young; it is you who must surrender.

And I didn't reply to my brother's letter. Let my father be free with his own attitude. Especially too as I didn't really know my father well. Since I was little I had lived with grandfather, so father was really no more than a title to me. Every time I met him, all he wanted was for his authority as a father to be acknowledged. It was up to him! I had no business with his anger and his attitude. If Father withdraws me from H.B.S., that too is his right. And a Native only got into school if someone with position guaranteed him. Only it was not Father who guaranteed me, but Grandfather. And it was not certain that the school director would accept Father's request. If he did, so be it. I now felt that I had accumulated enough means to study by myself, to enter the world walking on my own two feet.

Four days after Mr. Mellema's corpse was found he was buried at the European cemetery in Peneleh. We all attended. Most of those who attended were inhabitants of the business's villages. Seven reporters also witnessed the event, along with Dr. Martinet, Jean Marais, and Telinga. The burial was organized by the Verbrugge Burial Company.

Dr. Martinet took the job of representing the Mellema family. During the burial ceremony he told of his great sympathy for the Mellema family, especially Nyai Ontosoroh and Annelies, all of whom had been put through such trials over the last five years. Only a person who was truly strong could bear them. And the person involved was a Native, too, who was aided only by her clever and adroit daughter. And those trials weren't over yet, because the matter still had to come to court.

His pronouncements, expressed as they were as sympathy, were soon reproduced in the colonial press, both Malay and Dutch. Dr. Martinet became a target for journalists who demanded an explanation of his speech. He, who understood that such an explanation would be turned into a sensational serialized story, remained unrelentingly silent. So the Dutch-language press in their own way and style rejected Dr. Martinet's sympathy, which was directed at one who was only a Native woman, and a concubine too, who perhaps was not even clear of any wrong-doing in the case. There have been many proven cases of nyais conspiring with outsiders to murder their masters. The motives: lust and wealth. In the nineteenth century alone, there could be listed at least five nyais who had gone to the gallows. Even the character Nyai Dasima, of the popular Malay novel, could have carried out the same crime, had not her master, Edward Williams, been such a wise person. But even her story ended with a killing. Only it wasn't Edward Williams who was the victim—but Dasima herself. The paper closed its piece with the suggestion that Nyai Ontosoroh be investigated more thoroughly. Meanwhile a Betawi paper suggested that this person Minke was a character who should be more thoroughly investigated.

Dr. Martinet and Maarten Nijman collected a great many newspapers from other towns and passed them on to us.

After reading all their comments and proposals, Mama stated:

"They can't stand seeing Natives not being trodden under their feet. Natives must always be in the wrong, Europeans must be innocent, so therefore Natives must be wrong to start with. To be born a Native is to be in the wrong. We're facing a more difficult situation now, Minke, my son. [That was the first time she called me 'my son,' and tears came to my eyes when I heard it.] Will you run from us, child?"

"No, Mama. We'll face it together. We too have friends. And, I ask Mama, don't think of this Minke here as a criminal."

"They have all the means they need to make us scapegoats. But while none of us have been arrested—especially Darsam—it means the police haven't been influenced."

Another article, obviously written by Robert Suurhof, accused me of being an unashamed sponger, sucking up other people's wealth and representing myself to the public as a "church-bird-without-sin"; but I was actually someone without a family name, without anything. My only capital, it said, was my crocodile daring.

The paper wasn't, of course, the *S.N. v/d D.*, but a daily famous for its addiction to scandals and sensations in all fields, with staff who were sensation maniacs. Or as Dr. Martinet put it: sick people, like Titus in Roman times. Dr. Martinet visited us to express his solidarity.

"Don't drown."

No matter how one humored oneself, no matter what salve one applied to one's heart, Suurhof's article struck hard. The pain was felt even in the hairs on my neck.

"I'll take him to court, Mama."

"No!" forbade Nyai. "You'll never win."

"If Mama refuses to confirm what he says, I'll have already won."

"Mama is on your side," said the woman. "But you'll never win if you take it before the law. You'd be facing a European, Nyo. The prosecutor and judge will do you in and you don't have any court experience. Not all attorneys and barristers can be trusted, especially where the case is one of a Native suing a European. Answer that article with another of your own. Challenge him with words."

This person who says he knows me is perhaps my friend; a good friend or a bad friend, I answered in my article. Why doesn't he show his face in the open, why does he prefer to hide behind a mask when he launches his filth? Come out in the open, sir, show your own face. Why are you ashamed of your own face, your own name, and your own deeds?

My article, which was first published by Maarten Nijman, was then circulated more widely through an auction paper that had been able to turn itself into a general daily as a result of the Herman Mellema affair, though most of the paper was still advertisements. In all of Surabaya there were six auction companies. Each had their own auction paper. Only one was able to turn itself into a proper daily newspaper.

How much have I stolen from the late Herman Mellema? Tell us, sir. Give it all in detail if you can. You can ask assistance from Herman Mellema's family. Hire an accountant if you like, I wrote.

Truly, I would never have guessed. The attacks on me came roaring in. Mama was right—and I hadn't even brought it to court. The controversy didn't focus on the truth or otherwise of the accusation that I was a sponger sucking on Herman Mellema's wealth. The burning issue shifted to color difference: European versus Native. Papers in the other towns started meddling in the affair. So for one month I had not a single opportunity to look at schoolwork. My daily business was to respond to people's ignorance. Maarten Nijman gave me every report attacking us, and I had to reply.

Miss Magda Peters also came to express her sympathy:

"Yes, this is how it is in all colonies: Asia, Africa, America, Australia. Everything that is not European, and especially if it is not colonial, is trodden upon, laughed at, humiliated, for no other reason than to prove the supremacy of Europe and of colonial might in every matter—not excluding ignorance. Don't forget, Minke, those who first came to the Indies were mere adventurers, people Europe itself had exiled. Here they try to be even more European. Trash."

We listened to her expressions of sympathy, and to the oaths, in silence.

We tried to ensure that Annelies was kept out of the affair. It seemed our efforts were fairly successful. In this way an alliance developed between Nyai and me as we confronted the world outside the house.

"If you agree to fight them at my side, Minke, child, Nyo, you fight them to the end. If they later find themselves cornered, be careful—they will gang up on you. It's happened so often before. Do you dare?"

"We will never rest, Mama, in dealing with this problem. I think Minke is not a criminal. I reckon, Mama, he won't run."

"Good. In that case you don't need to go back to school yet. This fight is more important than school. At school they will gang up on you and will hurt both your body and your feelings. By facing this situation now you will learn to defend yourself and to go on the attack in public, before all races. You will graduate with the diploma named fame."

Unexpectedly, there appeared in a European-owned Malay paper an article defending me, written by someone calling himself Kommer.

If Minke, alias Max Tollenaar, has actually and quite plainly broken the law, he wrote, why have none of his accusers taken him to court? Do they consider that the law in the Netherlands Indies doesn't fulfill their needs? Or are they deliberately insulting the law and exposing the impotence of our honorable law officers? Or do these not-so-honorable gentlemen want to create a new law of their own?

As a consequence of this article several legal people began arguing among themselves, and attention was shifted away from me. And that diploma called fame, promised by Nyai, did not come my way.

Nyai Ontosoroh seemed nonplussed in facing all the possibilities. In all this extraordinary business Annelies became even more absorbed in her work. Relations with the outside world were surrendered to Mama and me. And all of a sudden I was acknowledged as the only man in the house, though not formally, of course.

The court case couldn't be put off any longer. Robert Mellema and Fatso were still not to be found. So the court was to try Babah Ah Tjong as the accused. A white court, a European court! Not because Ah Tjong had *forum privilegiatum,* the right to be tried under European law, but because the victim was European, though the accused was not, something I found out later. He was accused of the premeditated murder of Herman Mellema, murder carried out both gradually and in a final act.

Perhaps it was the biggest court case ever in Surabaya. Aroused by the reports and the arguments in the newspapers, the inhabitants of Surbaya, of all races, flocked to witness it. It was also reported that many people came from other towns. Nyai's brother from Tulangan also came.

People said it was the most expensive court trial ever. No less than four sworn interpreters were used: for Javanese and Madurese, Chinese, Japanese, and Malay. All the interpreters were Pure-Blood Europeans.

Mr. Telinga, Jean Marais, and Kommer also came. Kommer said that for as long as he'd been a journalist this much-feared building had never experienced such a cheerful group of visitors.

An owner of an auction office and paper that I knew also attended.

The Surabaya H.B.S. closed for the first time in its history: The teachers and students shifted their class to the court-building compound.

Dr. Martinet was called as an expert medical witness.

Babah Ah Tjong hired a defense lawyer from Hong Kong who spoke in English. So they had to get another interpreter.

People said that this was also the first time a Chinese had been tried by a European court.

The trial passed quickly at first. Dutch was used. It was difficult to get from Ah Tjong a confession of the motive behind the murder, even though he did confess to the poisoning using a Chinese prescription unknown to the medical world. He wouldn't tell the formula. All he would admit was that it made the victim lose his sense of balance, as was proved at trials of ten murderers in Kalisosok jail.

At first Ah Tjong denied that the mixture could do any damage. Its only use was as an aromatic for palm wine, he said. A Chinese physician was called as a witness. He repudiated Ah Tjong's explanation and the accused was pressed on this, the weakest aspect of his defense, which brought him to an eventual confession of murder.

What was his motive?

At first, Ah Tjong said he was fed up with his customer who, after five years, still didn't want to leave. But he couldn't answer the question: Why was he fed up when all this time his customer returned him a profit? And why then was Robert Mellema taken in?

The questioning of Nyai Ontosoroh, who had become the star of the trial, made her go scarlet. She was not allowed to use Dutch and ordered to use Javanese. She refused, and used Malay. She explained that the late Herman Mellema's bill at Ah Tjong's was forty-five guilders a month. It was paid at her office to a messenger. Lately, she had received bills for Robert Mellema at sixty guilders a month.

Why did Robert have to pay so much?

Because, answered Ah Tjong, Sinyo Robert only wanted Maiko, who was the most expensive girl; and he wanted her just for himself.

Was it true that Maiko only served Robert Mellema? Maiko said no. She served whomever Ah Tjong told her to serve, including Ah Tjong himself. Especially as Robert Mellema had recently begun to lose his strength and his sexual desires.

To satisfy those interested, Maiko was questioned as to whether she had ever contracted a venereal disease while she had been a prostitute. The expert witness, Dr. Martinet, explained that it was true that Maiko had contracted syphilis.

Did not Maiko regret having spread such a sickness in another people's country? She answered that it was not her wish that she become ill. She did not make the sickness. Her task as a prostitute was only to serve the customers.

Still to satisfy those especially interested, another question was asked: Who gave the disease to you? With a clear and beautiful voice Maiko answered that she didn't know. If customers were infected because of me, it was not my fault.

Had Babah Ah Tjong ever expressed his dissatisfaction to Nyai? Nyai answered that she had never even met her neighbor. She had only met his bills. Their first meeting was in this court-room.

Finally, the court ran into many issues that could not be cleared up and which therefore annoyed many people. The absence of Robert Mellema and Fatso was an obstacle that could not be overcome. But of all the questioning I thought out of order, the worst was that about my relationship with Annelies. It made people laugh and giggle; and both the prosecutor and judge, each in their turn, could not let pass the opportunity to ridicule our relationship in public. My relationship with Nyai was also subjected to disgusting and uncivilized insinuating questions. I was amazed that Europeans, my teachers, my civilizers, could behave in such a way.

It was fortunate that the questioning did not become too involved, though I knew that the intention was to prove whether or not sexual relations had occurred between us, in order to use any such confession as the link to connect us with the murder.

Ah Tjong made things go lighter for us with his statement that Nyai and I, as well as Annelies, had no connection with the murder. And that statement freed us from further involvement in the case.

The trial went on for two weeks. The motive for the murder still eluded the prosecutor. The judge decided to postpone his judgment. The prosecutor was ordered to find Robert Mellema, the latter to be detained and questioned. The court's decision seemed to disappoint many people. Many people, so it appeared, expected the judge to bring down a death sentence because an Oriental had carried out the premeditated murder of a European. The judge ordered that Ah Tjong be kept in temporary custody.

His helpers received sentences between three and five years each. Maiko was ordered to be treated at a hospital under the care of a doctor, to be paid for by Ah Tjong, as he was her employer. Meanwhile everyone waited impatiently for Fatso and Robert to be caught.

17

The trial came to a temporary halt. I went back to school.

Everyone had collected in the schoolyard by the time my buggy stopped in front of the main gate. They put off their other activities just to take a look and stare at me as I passed.

Even before I had gone into class I was given a message from the school director. And so I reported to him. These were his words:

"Minke, both as an individual and as representative of all the school's teachers and students I would like to congratulate you on your victory in court. I would also like personally to congratulate you on your tenacity in defending yourself from public attack. I, and all of us, are proud to have a pupil as talented as you. The court trial was followed by all the students and teachers. You no doubt already know that. You have been the focus of much attention, because you are a pupil at this school. Now I'd like you to listen to the decision of the Teachers' Council that has come out of its meetings and its rather difficult discussions about you. Based on your answers in court—I mean those concerning your relations with Annelies Mellema—the Teachers' Council has decided that you are too adult to mix with your fellow students, and in particular that you are a danger to the female students. The Teachers' Council meeting does not dare accept the responsibility of answering for the safety of the female pupils to their parents and guardians. Do you understand?"

"More than understand."

"A great pity. A few more months and you would have graduated."

"So be it. It's all up to you, Director, to decide."

He put out his hand to me and said:

"Failure in school, Minke, but success in love and life."

By the time I left the office, class had begun. I could see all the eyes directed at me through the window. I waved and they waved back. It was that response that suddenly made my heart sad at having to be parted from all these people who, it seemed, still did care about this Native, Minke.

The buggy and its driver were waiting outside. I climbed aboard quickly. As the buggy started to move, I ordered the driver to stop. Someone was running after me, calling out. Miss Magda Peters. And I climbed down.

"A pity, Minke. I was unable to defend you successfully. I fought as hard as I could. It was impudent of the court to ask you about such private matters in public."

"Thank you, miss."

She went. I climbed aboard and, at my request, the buggy set off slowly. Yes, the court was indeed impudent. The prosecutor deliberately wanted to turn our lives inside out in public—a desire that was a kind of extension of Robert Suurhof's feelings.

As if repeating Dr. Martinet's question, the prosecutor asked in Dutch, which was then translated into Javanese: "In which room do you sleep, Minke?" And indeed I refused to answer that malicious question. But with the speed of lightning the question was directed at Annelies and spoken directly in Dutch: "With whom does Miss Annelies Mellema sleep?" And Annelies had no power to refuse to answer. So humiliating giggling and laughter was heard in the courtroom, quite loudly too.

The next question was flung at Nyai Ontosoroh: Nyai Ontosoroh, alias Sanikem, concubine of the late Mr. Herman Mellema: "How could Nyai allow such improper relations between Nyai's guest and Nyai's child?"

The surging laughter became more exuberant, more insulting, more demonstrative. The prosecutor, and the judge too, both smiled, pleased that they had been able to engage in the torment of the spirit of this Native woman, a woman envied by so many Pure and Indo of her own sex.

With a clear voice and in flawless Dutch—defying the judicial order that she use Javanese, and ignoring the pounding of the gavel—like the flood waters released from the grip of a hurricane, she began: "Honorable Judge, Honorable Prosecutor, seeing that you have already begun to make public my family affairs [hammer of the gavel; a reminder to answer questions directly], I, Nyai Ontosoroh alias Sanikem, concubine of the late Mr. Herman Mellema, look upon the relations between my daughter and my guest in a different light. I, Sanikem, am only a concubine. Out of my concubinage my daughter Annelies was born. Nobody ever challenged my relationship with Herman Mellema. Why? For the simple reason he was a Pure-Blooded European. But now people are trying to make an issue of Mr. Minke's relationship with Annelies. Why? Only because Mr. Minke is a Native? Why then isn't something said about the parents of all Indos? Between Mr. Mellema and me there were only the ties of slavery and they were never challenged by the law. Between Mr. Minke and my daughter there is a mutual and pure love. Indeed there are no legal ties between them. But when my children were born without any such ties, no one was heard objecting. Europeans are able to purchase Native women just as I was purchased. Are such purchases truer than pure love? If Europeans can act in these ways because of their superior wealth and power, why is it that a Native must become the target of scorn and insults because of pure love?"

There was turmoil in the courtroom. Nyai kept on speaking, paying no heed to the judge's gavel. She was forced to admit that Annelies was not a Native, but an Indo. And the prosecutor's voice thundered furiously: "She is an Indo, an Indo, she's above you! Minke is a Native, though with *forum privilegiatum,* the right to appear before this court, meaning he's above you, Nyai, but his *forum* can be canceled at a moment's notice. But Miss Annelies remains above Natives forever."

"Annelies, my daughter, sirs, is only an Indo, so is that why she may not do the things her father did? It was I who gave birth to her, who reared her, who educated her without a single cent of aid from you honorable gentlemen. Or perhaps it wasn't I who have been responsible for her all this time? You gentlemen have never worked for and worried after her. Why all the fuss now?"

Nyai no longer heeded the court's authority. A police officer was ordered to remove her from the courtroom. She was dragged from her place, unable to resist. But her tongue did not stop letting fly words, bullets of revenge:

"Who turned me into a concubine? Who turned us all into nyais? European gentlemen, made masters. Why in these official forums are we laughed at? Humiliated? Or is it that you gentlemen want my daughter to become a concubine too?"

Her voice rang throughout the building. And all present were silenced. The police officer dragging her away moved faster to finish the job. She, this Native woman, had now become the unofficial prosecutor, plaintiff against the European race—a race now ridiculing their own deeds.

She went on speaking all the time they were dragging her away.

And now the buggy traveled slowly along roads already showing signs of the early morning traffic. And at the school court too the gavel had struck: I was no longer the same as my fellow students. I was a danger to the girls, dismissed dishonorably from school. If, for example, the secrets of the teachers were to be exposed in public, cut open without mercy, who could guarantee they wouldn't turn out to be more corrupt? Doesn't everyone have their personal secrets, which they carry with them until death? And that prosecutor and that judge, both of whom showed me no mercy—who knows, might they themselves be keeping concubines, openly or secretly? Perhaps if there were no public or legal controls, their behavior might be even more rotten than that of Herman Mellema towards Sanikem?

Traveling along in the buggy I felt that every person I saw was pointing accusingly at me: That is Minke, who sleeps in the same room as Annelies, a woman he hasn't yet married.

That is a Minke now different from all his friends, different from everybody else—his situation was exposed in court, wasn't it? While the others weren't? And the judge and prosecutor didn't expose themselves either?

What I was feeling then, such very depressed feelings, my ancestors called *nelangsa*—feeling completely alone, still living among one's fellows but no longer the same; the heat of the sun is borne by all, but the heat in one's heart is borne alone. The only way to obtain relief was communion with the hearts of those of a similar fate, similar values, similar ties, with the same burdens: Nyai Ontosoroh, Annelies, Jean Marais, Darsam.

So I went to Jean's house.

"You're dispirited Minke. Expelled from school? Chin up!"

And he whose chin was always buried deep could now say chin up! It felt as if all things joyous had been eliminated from my heart.

"Your school is too small for you now, Minke. If Minke has been broken like this, there's still Max Tollenaar, isn't there?"

He looked at me as if I had some secret store of spirit. He didn't realize that my humiliation meant it would now be more difficult to find orders. I told him. He was silent for a moment. Suddenly he burst out laughing. And I was somewhat hurt.

"Do you know, Minke, I see a joke in all this."

"There's nothing funny," I said, annoyed.

"There is. Do you know what? There is only one medicine that can cure you. Get married, Minke. You must marry Annelies. Show to the world that you're not afraid of confronting even the eye of Satan. You'll become like the others. They're not asking much, only that you return to their fold—stupid, uncultured people. Marry, Minke, just marry."

"Magda Peters said she thought the court was unjust towards us, even insolent."

"Yes, it was uncivilized. That's the most apt way of describing it. There are some Malay-Dutch papers that have said the same thing. Only not as strongly as that. Those sorts of questions should only be asked in closed court."

"Yes. But there was one Dutch paper that called Mama insolent, said that she created turmoil in the court. But they didn't even print her words."

"Read Kommer's article. His anger was like that of a wounded lion. He's on your side."

"Tell me. I don't feel like reading."

"He wrote that the actions of the judge and prosecutor were insults to all Indo-Europeans born out of concubinage, out of a relationship with a nyai. Their children, if acknowledged by the father, are not considered Natives. If the father doesn't acknowledge them, they become Natives. It means: Natives are the equivalent of children born of a concubine whose father won't acknowledge them. He also condemned the court's exposure of your private affairs. Kommer said that the prosecutor and judge did not have European morals; it was worse than the Native court set up by Wiroguno to try Pronocitro almost two hundred and fifty years ago. Who were they, Minke? I don't know."

"I'll tell you another time," I said, and left.

As soon as I arrived home, I went straight into the office to report on the new disaster.

"Mama, what do you think about the idea of Annelies and me marrying?"

"Wait a while. What's the hurry?"

I told her about the troubles that had befallen my attempts to obtain orders. Troubles that might also befall Jean Marais.

"What can one do, child? Regrets don't achieve anything. The days at court have brought considerable losses to the business, and my position could deteriorate to that of any ordinary nyai, to be humiliated in public, to be sneered at out of the corners of people's eyes. We have to make up those losses first, child. Because without this company doing well, this family will lose its honor. I hope you can understand."

I observed Nyai's lips as they spoke so calmly. She hoped so much for my understanding.

"Minke, I have reflected on the strangeness of life for a long time now. If I can't save this business, my position will fall to that of any ordinary nyai. Annelies would suffer greatly. I will have been a complete loss as a mother. She must be respected more than an ordinary Indo. She must become a Native honored among her own people. Such honor and respect can only be obtained through this business. It's strange, child, but that is what the world demands."

Annelies herself was working out at the back.

As I sat on the chair in the office, the question of Pures, Indos, and Natives hovered before my mind's eye, clearing away that humiliating self-pity. Everything formed a network like that of a spider's web. And in the middle of the web were the concubines and nyais. They don't catch all the victims that come to them. On the contrary, the net catches up all possible humiliations that they then must swallow. They aren't employers even though they live together in the same room with their masters. They are not included in the same class as the children they themselves have borne. They are not Pure, not Indo, and can even be said not to be Native. They are secret mountains.

And my hand moved fluently across the page. Kommer's ideas were the backbone of my article this time. And the sun set. And my article began to take form.

Ya Allah, even out of such humiliating self-pity can come greater understanding about Your people. It was You too who ordered humanity to form nations and multiply. You have blessed relationships between men and women of different social and economic levels. Why have You not blessed this relationship between people of the same social and economic level and formed of their own free will—only because it has not been carried out according to Your rules? And You have allowed all this to occur, so as to give birth to the Indos, who have so much power over those born with Your blessing.

I turn to You now because those nearest to You will not answer me. You must answer now. I am writing only what I know and what I think I know. Does not all knowledge and learning originate, in the end, from You Yourself?

Ten days after Max Tollenaar's article about the issue of Pures, Indos, and Natives was published, Magda Peters came to Mama's house during school hours. The school director wanted to see me. And I refused to go on the grounds that I no longer had any connection with the school.

Nyai also objected to my going.

Annelies ran away into her room.

"Something has happened," said our guest. "No matter how you feel, you must come. But first of all accept my congratulations. Your last article was a true call to humanity, a powerful incentive to people to think more wisely. And you're still so young. . . ."

So in the end I went.

All along the journey, Magda Peters twittered about how good it was to have a student of whom she could be so proud. After all my recent experiences I felt soothed by this.

The director received me with a friendly smile. All the students were given the rest of the day off. All the teachers were called together. A kangaroo court? Why was all this being done just for one person? Why was I so important?

The director opened the meeting.

"It has now become a European tradition to make judgments on people based only on their achievements. Even on this spot of land called Surabaya that tradition must be maintained. We are not going to ask: What is this man like? No, because that is a private affair. He is valued because of his achievements, because of what he has contributed to his fellow human beings."

And this beginning soon brought him to my latest article.

"Moving. Touching upon our sense of sanity. More than that: true. It seems that the humanist conscience of Europe, for so long absent among the Indies Natives, has begun to grow within Max Tollenaar, one of our own students . . . Minke."

I didn't understand at all what was meant by European humanism.

"There have already been seven letters, two from graduates, which protested against our decision to expel Minke. One said this person must be helped, not expelled, even if it means taking some kind of special measure. The assistant resident of B—— even felt it necessary to come to Surabaya to see the resident of Surabaya to discuss the matter. The resident had no view on the matter, but the assistant resident offered to be Minke's guardian while he was at H.B.S. He was even going to seek a meeting with the director of the department of teaching and religion if his efforts in Surabaya were not successful.

"So, for the first time, one of our decisions is being tested and challenged, though it is not because of those tests and challenges that we must review our decision, but because of what we call our European humanistic conscience, in the name of our ancestors, and European civilization today.

"Now, here is Minke, Max Tollenaar, standing before this respected council of teachers. This council will review its earlier decision and decide on some new policy."

Like a lioness who has lost her child, Magda Peters roared,

clawed, attacked in the interest of her lost child. Her freckles stood out more sharply. Her eyes blinked more quickly. Finally, in a low voice, slow and halting, she closed with:

"Education and teaching are nothing if not works of humanity. If someone outside school has developed into an individual with a sense of humanity, like Minke, we should be grateful and thank God, even though our part in the forming of this individual is so tiny. Such an outstanding individual can only be born out of extraordinary conditions and circumstances, as is the case with Minke. So I propose: Let us accept him back in school so that he can be given a stronger foundation on which he may build his future."

The council continued, with myself as the mute accused who did not know why he was being obliged to witness it all, and finally it was decided that I would be accepted back as a student. But with special conditions: I must sit at a special desk set apart from the others, and whether in class or out, I must not talk to fellow pupils, either to ask a question or to answer one.

"What is your opinion, Minke, now that you've heard all this?" asked the director, who appeared to want to wash his hands of any past sins.

"While ever there is the possibility, I will continue my schooling as I indeed originally desired. If the door is open to me, I will certainly enter. If it's closed to me, I have no objections either if I do not enter. Thank you for all your efforts."

The meeting closed. With dark faces, except for Magda Peters, all the teachers shook hands to congratulate me. My Dutch language and literature teacher was so satisfied with herself and considered all that had occurred her own victory!

As a closing ceremony the school director handed me unstamped letters from Miriam and Sarah de la Croix.

The school was still. The H.B.S. building, compound, even the gravel all seemed foreign to me as if this were the first time I had seen the school. The teachers' looks could be felt tickling my back. I walked straight to the buggy without turning to look back again.

"Go slowly," I ordered the driver, Marjuki, in Javanese. "Straight to the newspaper office."

Halfway there the driver said shyly:

"Master looks so pale and thin."

293

"Yes."

"Why don't you take a holiday, seek a cure, Master?"

"Yes, later, when I've graduated from school."

"Three more months, Master?"

"Yes. Still three more months."

"What's the use of school, master, if you already have enough of everything?"

"Yes, what is the use? But if I don't succeed in school, Juki, I'll feel like I won't succeed in anything else later."

"Master has already succeeded in everything."

"Succeeded, how?"

"That's what people say, only what people say, Master. Noni . . . wealth, cleverness, you know important people, Dutchmen, not just anybody. . . ."

"Is that what people say?"

"Yes, Master, and so young, handsome; and in a little while you'll become a bupati."

"Ah, forget it, Juki, forget it."

At the office of *S.N. v/d D.,* Maarten Nijman offered me a full-time job with his paper if I was expelled from school. The work would be very interesting, he said, even if the wages somewhat low, only twelve and a half guilders. I told him all about the teachers' council meeting and the decision it had just taken.

"So Miss Magda Peters defended you with such great spirit? Ah yes, Magda Peters. Are you close to her?"

"The wisest of my teachers, sir."

"Hmm. I think it would be wise if you distanced yourself from her somewhat."

"She is so kind."

"Kind? Yes, that's her way of leading people astray, I think."

"Leading people astray?"

"You must have heard at one time or other: People can be led astray with kindness too."

"Led astray how?" I asked, amazed.

"She is a fanatical radical. She's one of those busy with the 'Indies for the Indies' movement. Have you heard of it?" I shook my head. "They say the Indies should be equal with the Netherlands. She and her kind don't want to know about all the limitations that exist in the Indies. Only disaster will befall those who

294

dare fight against, let alone defy, those limitations. And among all those limitations the most numerous are the unwritten limitations. In the Netherlands, of course, there is total freedom. Here no such thing exists. There is nothing wrong with being a liberal as long as the limitations here are recognized and no one causes any commotions. That's something you should know. It's fortunate that no Natives have joined that movement. Imagine if you had joined it! If a liberal is condemned by the government—no matter what he did wrong—if he's a Pure-Blood, exile from the Indies would be his most severe punishment. If he was an Indo, punishment would be more bitter: dismissal. If a Native, I think that he'd lose his freedom altogether; he'd be locked away without any trial— because there's no law dealing specifically with this sort of thing. Be careful you don't end up the one who gets in trouble. Your country is not the Netherlands, not Europe. If you get in trouble no one among the liberals will be able or willing to help you."

"She's my teacher, Mr. Nijman, my teacher."

"Look, Mr. Minke, the Netherlands Indies runs on rumors. And it's worth listening to those that come down from above. There have already been rumors about Miss Magda Peters. You've already experienced a lot of trouble lately. Don't add to it, Mr. Minke."

He spoke for a long time about the activities of the liberals, politely, but in a tone that rejected and criticized them. At one stage, he accused them of wanting to overturn the situation in the Indies which, he said, was already consolidated, orderly, secure, tranquil and where its people were protected as each day they went about making a living.

"And, Mr. Minke, under the Native kings, your people were never secure, never at peace; there was no legal protection, because indeed there was no law. What hasn't the Netherlands Indies government done for the people? The liberals indeed have strange ideas about the Indies."

"But they're Europeans too," I said.

On the way home in the buggy, it kept occurring to me how all these conflicts made the situation so complicated. Now we had to add to all this: Pure against Pure. And there was still the position of the Orientals, while Maarten Nijman also wanted humanism, but rejected liberalism. It was turning out that the more one mixed with people the more often different types of issues emerged, ones that I had never dreamed existed, and they were popping up like mushrooms.

Nijman had warned me to be prepared for the present and the future. And, he said, it could be that in the near future Magda Peters might have to leave the Indies. It was not only possible but probable. The rumors were rife and that was taken as an omen. Before such an event actually took place it would be best if I distanced myself from Magda Peters, he said. "Magda Peters may only be ordered to leave the Indies, but you could find yourself in a place that you would never be able to leave."

Nijman didn't want to explain the limitations he talked about. Good. I would ask whoever was prepared to answer. Perhaps his words contained some truth, if there were such limitations, and they were real.

At the Telinga's house there was a letter from Mother and, as was usual, it was written in Javanese language and script.

Gus, everybody has felt both pain and sympathy as they followed your affairs in the papers. You are my manly son. That is all that supports me. As for your own affairs, you yourself must resolve them all. Don't forget what your Mother has said before: Don't run! Resolve your affairs well. You remember? If you ever run away from something, your schooling and your education will have been in vain, because my son would then be only a criminal. You are fond of the daughter of this Nyai Ontosoroh? That's up to you. I only say: Don't run from your own problems, because to resolve them is your right as a man. Seize the beautiful flowers, because they are there for him who is manly. And don't become a criminal in affairs of love either—one who conquers a woman with the jingle of coins, the sparkle of wealth and rank. Such a man is also a criminal, while the woman is a prostitute.

I hear too from those who read the Dutch papers that you've become a man of letters. Oh, Gus, why do you compose in a language that your mother cannot understand? Write the story of your love in the poetry of

your ancestors so that your mother and the whole country may sing them.

Don't worry about your father, he has a poem of his
own.

Ah, beloved Mother. How great a love from me do you deserve!
You have never punished me, never passed judgment on this son
of yours. Since I was little you have never even pinched me. Now
you find no error in my relationship with Annelies. You seek of
me that I write in Javanese, a language that you can pronounce
with your own tongue. How I have disappointed you, Mother,
not being able to write Javanese poems. The rhythm of my life
writhes so wildly it could never be forced into the poetry of my
ancestors.

My communion with Mother was destroyed by Mrs. Telinga
with her usual nagging:

"What about this, Young Master, there may be no shopping
done tomorrow," and that meant that I'd have to produce at least
one talen from my pocket.

At Jean Marais's house I found May asleep on a bed, now
equipped with a new mattress but still no sheets. Jean himself was
daydreaming. The workshop at the back of the house was rather
quiet.

"Jean, tomorrow you can begin painting Mama. It's best if
you paint while she's doing her correspondence in the office. I'm
back at school beginning tomorrow. And May can stay at Wonokromo while you're painting, Jean."

"I'll come, Minke." His voice still sounded lonely. "I don't
really feel like painting now."

"It was you that wanted to do it before."

"She's so strong, Minke. Her personality is so strong. I admire her, no more so than during the court case. She's such a
determined person, with such vision. I could drown before her."

What was he trying to say: that he had fallen in love with
Mama? Only that he had no way of communicating it?

The Frenchman didn't say anything more.

"Have you ever suffered because of love, Jean."

He raised his head and smiled. He asked in return:

"Have you ever heard the life story of the great French painter

297

Toulouse-Lautrec? His immortal paintings now hang on the walls of the Louvre?"

"Of course not."

"He achieved everything in life."

"Why, Jean."

He smiled mysteriously and wouldn't say.

Still yawning, May climbed up into my lap.

"Have a bath, May. Let's go to Wonokromo. Tomorrow you can come with me on the way to school."

"By buggy from Wonokromo?" she asked, her eyes gazing at her father.

Jean Marais nodded in confirmation.

"You too, Jean. No need to wait for tomorrow. Let's go now."

The three of us departed. The buggy was very crammed.

And in the evening, with Jean Marais as a witness, it was decided: Annelies and I were to marry as soon as I passed the H.B.S. exams.

The world and my heart greeted each other in peace.

18

The graduation party was also a party within a party.

For three months I studied and studied and that's all I did. I didn't do any work for Jean. Studied and studied. And in the meantime my life had been restored to something like it was earlier.

I would be allowed to socialize once again with the other students at the graduation ceremony. I would be a real part of the student body again—even though only briefly. We would all soon be embarking on our separate journeys into that unlimited life before us.

The parents and guardians sat in rows, all of them: Pures, Indos, several Chinese, and not a single Native.

Mama refused to attend, so I came with Annelies. And it was the first time she had left the house to attend a party. She wore her favorite black velvet dress with a three-stringed pearl necklace, a brilliant bejewelled medallion, and bracelets. And there could be no doubt: She now rivaled the Queen in both natural beauty and her adorned appearance.

Like all the other students who were to receive their diplomas, I wore white clothes like the civil servants, except without the yellow buttons with the letter *W* for Wilhelmina on them.

The two of us entered the hall, where we were greeted by Magda Peters dressed in her formal clothes. She greeted Annelies with great enthusiasm.

"Prima donna! You are the queen of the party!"

With all eyes on her, Annelies did not refuse to be escorted to sit amongst the audience. Both girls and boys turned around to see my queen. Now they knew: The world had become my kingdom, and I had seized it without a duel. I sought out Robert Suurhof so he would have no chance to hide his face. Instead it was Jan Dapperste who came into view, waving his hand. I replied with a nod.

Sitting in the chair there, I remembered Mother. How glorious it would have been if she could have witnessed her son accepting his H.B.S. diploma. That noble woman was not present. And I felt an emptiness in the merriment and the grandness of the occasion.

The hubbub turned to silence. "Wilhelmus" boomed out as everyone joined in singing under the witness of the Tricolors: flags and ribbons. Then the director spoke briefly. He congratulated those who had graduated and wished them well as they set off on their brilliant careers, and he prayed for the greatest success for everyone. To those who would continue their studies in the Netherlands at university, he wished them happy sailing and prayed that they would become good scholars, of use to the Netherlands and the Indies and the World.

The European inspector of teaching did not speak.

Then the program moved on to calling up those who had passed the 1899 state exams. The teachers were standing in a row behind the director.

Stillness and tension.

"At the close of this school year, approaching the close of the nineteenth century, of the forty-five pupils who sat for the state exams in the Indies, the one who came first was from H.B.S. Batavia. Eleven failed and have been asked to repeat. The graduate with the second best score is from Surabaya, which means he was first in Surabaya."

Everyone cheered loyally.

I guessed that each student's heart was palpitating as he imagined himself as number two in all the Indies and number one in Surabaya. Even I had dreamed of it.

"Graduating second in all of the Indies, first in Surabaya, the student's name is . . . Min-ke."

I shook. I had no idea. In fact no one ever imagined that a Native could beat Europeans. Such an idea was taboo in the Indies.

"Minke!" called the director.

I still couldn't stand. My fellow students on each side of me forced themselves to help me get up.

"Minke!" called out Magda Peters, waving.

And so I stood up; my legs were still unsteady. Everyone would no doubt witness this pathetic condition I was in. There was no more applause supporting me, only because the person called up was a Native. Even the teachers didn't applaud. Then came some weak clapping. It was easy to guess who: Miss Magda Peters. Perhaps even Annelies didn't clap. She had never participated in this type of gathering before. She was probably sitting there completely dumbfounded—that child who'd never mixed socially—like the child of a mountain.

I went up on the stage and received the diploma and the congratulations. My hand shook visibly.

"Steady, Minke," the director whispered.

Slowly I walked back to my seat, accompanied by the weak clapping of the teachers, and then that of a few students, and then of some of the audience.

Number five after me was Robert Suurhof. Last was Jan Dapperste. When Jan returned to his seat, there appeared from amongst the audience Preacher Dapperste, a Pure, who greeted Jan with an intimate embrace. The preacher's wife too. If Annelies had understood what this was all about, she would have done likewise. She didn't.

The party began. First and second class were to put on a Bible play; it was called *David and Bathsheba,* and was produced by one of the teachers.

Audience and students sat together now. Annelies was beside me.

Before the play began, the director came up to me and gave me a telegram from B——: Congratulations on passing the state exam second in the Indies, from Miriam, Sarah, and Herbert de la Croix. It seemed that they knew my results before me, the person it affected. The director shook hands with Annelies and was very friendly. Even so I was tense and anxious, afraid that he still might let fly some insult, openly or not. But no, he didn't insult her. It seemed he greeted her sincerely.

"Would sir accept our invitation to you, the other teachers, and the students to attend our wedding party next Wednesday? At seven in the evening?"

"So fast?" Once again he shook hands with us.

Annelies responded to his congratulations coldly. And if I recalled Dr. Martinet's explanation, I could understand why.

He shook me vigorously by the hand and then clapped merrily so that people looked around at us.

"May I announce it in a moment?"

"Thank you, sir, of course, as an official spoken invitation."

"Why no printed invitations?"

"Well, sir, after our past experiences . . ."

Magda Peters, who sat listening, also shook hands with us, without comment. I don't know what she was thinking. At the very least her eyes weren't blinking fast.

The director went again. It was announced that scene one was about to begin. Slowly the curtain rose. Spread out before us was a stony landscape where later (perhaps) Bathsheba would bathe and where David would glimpse her body. But Bathsheba still didn't appear even though the curtain was fully raised. Nor the Prophet David. People began to stretch their necks forward, looking for the beautiful Bathsheba. Instead, out of the stony landscape there appeared the director, smiling and taking off his bow tie.

The whole gathering burst into hearty laughter. The director couldn't help but burst into a big grin also. So this unrobed, unturbanned David (but one who did wear a bow tie) apologized, but he said there was something he had to do. It would diminish his announcement's significance if it was done after the performance. Then he announced our invitation.

"Invitees besides teachers, students, and graduates? There aren't any."

There was scattered laughter.

"On behalf of all those who may not be able to attend, perhaps because they'll be returning quickly to their own regions or because they already have plans, as director of the Surabaya H.B.S. I wish to congratulate the bride- and groom-to-be and pray that they will live happily forever. Thank you."

And he descended from the stage passing Bathsheba, who was sneaking a look from behind the curtain.

Our wedding party, which we had planned to be a simple affair, turned into something much grander as a result of the sudden invitation announced at the graduation. Nyai approved. She was very happy to hear Annelies's report of how it was announced.

"This party will also celebrate your victory in the examination, child. Despite facing so many trials, you still passed brilliantly. You overcame all trials."

A few days before the ceremony, Mother arrived. She was the only representative of my family. Nyai greeted her joyfully, as if they were old friends. Mother quickly came to love Annelies, her future daughter-in-law. It was as if she could not bear to be far from her and never grew tired of staring in admiration at her beauty.

"Ya, Sis," she said to Nyai, the future mother-in-law of her son, "a child so beautiful, like Nawangwulan. Perhaps even more beautiful than Banowati. *Ya Allah,* Sis, I never guessed, I never thought that Sis would take my son as your son-in-law. Neither in this world or the next will I ever forget it, Sis."

"Yes, Sis they love each other. Only I ask your forgiveness because my child has no race, stems from a . . ."

"Ah, Sis, if a girl is as beautiful as this, she already has everything."

In the evening Mother whispered to me:

"Gus, you're truly fortunate to have obtained a wife so beautiful. In your ancestors' time a woman as beautiful as that would spark a great war."

"Does Mother think I did not go to war to win her?"

"Ah, yes, yes . . . you're right, Gus, and of course your victory was glorious."

We were married according to Islam. Darsam acted as witness and as guardian to Annelies as stipulated in Islamic Law. It took place at exactly nine o'clock in the morning. As was the custom, and with feelings of gratitude, we both knelt and made obeisance to Mother and Mama.

Tears poured forth from both of them as they accepted our obeisance and they blessed us with halting words. And Annelies cried also. Perhaps she felt there was something missing because there was no father present to share in the happiness of the day. Perhaps.

Mother and Mama put their arms around each other's shoulders, gazed at each other with tear-filled eyes, and embraced. To be moved to tears that way is humanity's purest emotion. Such emotion also means pain, hurt in one's psyche, because people come face to face with their own birth as humans, naked and bare of all pretension and civilization.

A small feast followed; then afterwards the real party.

For the inhabitants of all the company villages, our marriage meant a big feast. The paddy-drying area was turned into a big covered pavilion. Everyone was given a holiday with full pay. The herdsmen who couldn't leave their work received triple pay. Five young calves were slaughtered. Three hundred chickens met their end. Two thousand and twenty-five eggs. The whole of the day's milk production was surrendered to the kitchen. Every company carriage, whether being used or not, was decorated with multi-colored paper.

Never had the inhabitants of Wonokromo witnessed so big a wedding party.

Annelies once told me: Mama will give me anything I ask for at this wedding party. And she said also: She wants to see as many people as possible around her child sharing in the joy. So she would never regret it all her life.

Neither Annelies nor Mama wanted any dowry. "What more do we wish for?" said Mama. Annelies had already got everything from her future husband. If there had to be a dowry, said Annelies, it is something I haven't yet got from him: his promise of faithfulness *while ever I live.* And I gave it to her during the marriage ceremony.

At five in the afternoon, there was a knock on my door. Jan Dapperste entered. He was well dressed in an old-fashioned style.

"Forgive me, Minke, if I've come too early. I've come early on purpose so I can help." He then sat down as if he hadn't been acquainted with a single chair for the last fifteen years.

In a complaining tone, he continued, "You're indeed a child

304

of May, you've got everything you've ever wanted. You succeed in everything you do. A few more years and you will be a bupati."

"You're talking some ill-starred child lamenting his fate."

"You're not wrong. I've run away from Papa and Mama. After the ship set sail for Europe, I jumped overboard and swam ashore. An adopted child who doesn't know how well off he is . . ."

"I've heard you curse yourself like that at least three times now."

"I'm sorry, especially on this, such a happy day for you. It's not proper. I'm sorry. Help me, Minke. I don't want to leave Java. I'm not Dutch, not Indo."

"I've heard that often."

"Ya. And there's more than that: I've never felt happy with the name Dapperste."

The Preacher Dapperste's family had no children. They had taken care of Jan since he was small, baptized him, and given him their family name, Dapperste. Since then he's been called Jan Dapperste. He didn't know what his name was before all that. The preacher had tried to adopt him through the courts. He never succeeded because Dutch law didn't recognize adoption. So his name was something recognized in the community, but not by the law.

"Since I was small I have always been a coward. You know that too. The name Dapperste—the courageous one—has always tormented me."

Yes, all our fellow school students knew. Some even changed Dapperste to Lafste—Jan de Lafste—the coward. And if his story was true, then for no other reason than to free himself of that name he had changed into a brave person indeed: diving into the sea and running away from his parents. I still didn't really believe him.

"So who are you living with now?" I asked.

"Here and there . . . I want to get a job here in Surabaya with my H.B.S. diploma. The only problem, Minke, is that the name Dapperste is on that diploma too. Must I go through life carrying that name to the very end?"

"You can change your name."

"Yes, I know. I've spent the last year seeking explanations of how to go about it."

"And?"

305

"You put forward a letter to the resident, who sends it on to the governor-general."

"Why haven't you done it then?"

He gazed at me with dull eyes, as if he weren't an H.B.S. graduate. He made a noise with his mouth and looked away.

"You can't write a letter? Come on, you can use other examples of official letters?"

"It's the duty-stamp, Minke; it's too expensive to get free of this name. Just the application alone costs one and a half guilders. For the letter of determination that I need there is another one-and-a-half-guilder stamp. I've thought and thought . . ."

"And why haven't you done it yet?"

"Come on, Minke, don't tell me you don't understand. Where would I get three guilders? And then there are still the postage stamps."

"Why didn't you just say you didn't know where you could get the money? Wouldn't have that been easier?"

"I'm sorry. Forgive me, Minke, it's so embarrassing speaking to you like this on this day of happiness for you."

"You don't regret my happiness."

"No, of course not. I thank God too with all sincerity and honesty."

"Then share too in my happiness."

"That's why I have come."

"Listen, Jan, Mama is going to expand the business. She's going to move into spices. You can try working there. You'd like that, yes? While waiting for the determination about your name to be issued?"

"Thank you, Minke. You've always been good and generous. It's a pity the governor-general's letter has to be preceded by an application—I haven't even made it out yet."

"The new company is being headed by an Indo, van Doornenbosch. I'll introduce you to him later. I'll look after everything myself."

He took my hand. His head was bowed down. He didn't speak.

"Don't just be silent. Talk with me while there's still some time."

"Thank you, Minke. That's not all. You yourself can guess

my situation. My accommodation, Minke, and the cost of traveling around Surabaya?"

Mother entered my room to prepare my adornment. That noble woman had struggled hard to make sure she would get this job. No one else would adorn her child, of whom she was so proud, now that he was to be a groom. In her right hand she carried a paper bag and in her left, a basket of flowers, some untied and some in bunches.

On seeing Jan Dapperste, who looked at her in a demeaning way, she hesitated.

"This is my Mother, Jan," I said.

Only then did my friend force a smile and bow in respect.

"Mother doesn't speak Dutch," I reminded him.

And Jan Dapperste began to speak in fluent High Javanese. I was quite dumbfounded to witness it. And I explained to Mother that he was a fellow graduate, the son of a preacher.

"The former picked-up son of a preacher," he corrected me.

"Child, Mother wishes to adorn her son now. Excuse us."

"Allow me to help, Mother."

"A thousand thanks, child. But no. This is the last job his mother will do for her child. I must do it myself. Do you think you could move to another place?"

Jan looked at me, his eyes shouting out for help. I knew he was tired. And more than that: hungry. I knew his behavior by heart. I took a piece of paper and wrote an order to Darsam to look after Jan.

"Find Darsam." He took the letter and went.

I could now light the gas lamp in my room, a sign that it was exactly six o'clock. The gas mains, which Darsam himself looked after, and which were located in a small stone house behind the main building, were now opened. The room lit up.

Mother scrubbed my face, neck, chest, and arms with a liquid whose name I didn't know.

"In the bygone ages," Mother began, just as she had when I was still small, "countries would wage all-out wars to win a maiden like my daughter-in-law, *mbedah praja, mboyong putri* was our ancestors proverb: Victory over kingdoms, possession of its princesses. Today things are more secure. It's not like it was when I was little, let alone when your grandfather was little. Even though the Dutch are so very powerful they have never stolen people's wives or daughters like the kings who ruled our ancestors. Ah, child, had you lived in those days you would have been constantly called to the battlefield to be able to keep possession of your wife, that angel. Maybe she's even more beautiful than that. Her cheeks, her lips, her forehead, her nose, even her ears—all are as if formed in wax, shaped according to all men's dreams. How proud I am to have her as my daughter-in-law, Gus. You have made me so happy."

"She, Mother, this daughter-in-law of Mother's, there's not enough Java in her."

"You are happy with her, aren't you? Be happy in the beginning, Gus, but then be ever vigilant—a child as beautiful as that . . . the gods will not be still."

Mother kept on looking to my body, and keep on talking too, and talking.

"You are lucky you don't have to be always fighting like your ancestors."

"Mother."

"Ah, if only I could take her to B——, Gus, everyone would come out of their houses to greet her. What about it? Will you both later come to B—— or not?"

"No, Mother."

"Ya, ya, I understand, Gus. So your mother must be the one to come here to see you, and my daughter-in-law, and my grandchildren."

"It would be Father who would object, Mother."

"Sst. Silent, you. So you've forbidden your wife to have her teeth filed? Don't her sharp teeth disgust you?"

"Let my wife's teeth remain as they were given to her, Mother."

"Like Dutch teeth, like the unfiled teeth of an ogre."

"Why are you scrubbing me like this, as if I've never bathed?"

"Hush! On this your wedding day I want to see you look like the child of the gods. So neither you nor I will look back on this day with any disappointment."

"What's the use of being the child of the gods?"

"Hush! It is not for yourself that you must be like the child of the gods. Today, all your ancestors will come to witness your wedding and give their blessings. I would never miss an opportunity to see one of my descendants. Imagine how I'd feel to see my son ascend the wedding throne looking like anything but a Javanese knight? And what will I say when I'm dead and see my grandchildren fail to be Javanese only because their parents did not look to things properly?"

"Do the ancestors of the Dutch also attend the wedding of their descendants?"

"Hush! Why are you concerned with the Dutch? You're still not Javanese enough yet. You don't obey your own ancestors enough. People say you've become a man of letters, but where are your poems that I can sing at night when I miss you?"

"I cannot write in Javanese, Mother."

"An, if you were Javanese, you would be able to write in Javanese. You write in Dutch, Gus, because you no longer want to be Javanese. You write for Dutch people. Why do you honor them so greatly? They drink and eat from the Javanese earth. You do not eat and drink from the Dutch earth. Why, why do you honor them so greatly?"

"Yes, Mother."

"What are you yessing? Your ancestors, the kings of Java, all wrote in Javanese. Are you perhaps ashamed of being Javanese? Ashamed that you're not Dutch?"

It would have been stupid to answer Mother's words, spoken so gently but containing an unanswerable harshness. Yes, everybody makes demands on me. Now Mother too. Mother knew and I knew I would not answer. She was speaking more to her ancestors, pleading that they forgive me, her favorite child. The ancestors must not be furious at me. Ah, Mother, my beloved Mother, a mother who has never tried to force her will upon me, has never hurt me, not even with only a little pinch, no, not with words, not with her fingers.

"Put on this batik kain. Now. Mother made this batik for you herself, and for this occasion. Four years I have stored it in a special box; every week I sprinkled it with jasmine flowers, Gus. After I heard people's stories about the newspaper reports of the trial, I sacralized the kain, Gus. One for you, and one for my daughter-in-law. Inspect your Mother's batik work and smell the aroma of the years and years of jasmine."

So I inspected it and I smelled it.

"Beautiful, Mother, wonderful. So sweet-smelling. And that

delicious aroma has been absorbed right down into the threads."

"Aah, what do you know about batik," and deliberately she didn't look at me, knowing that I'd be grimacing from the pain. "I dyed it red and blue with my own hands, Gus. And the dyes I made myself also. Smell its aroma again, the perfume of the dye is still there," and the kain was pushed under my nose.

"Delicious, Mother."

"Ah, you! I'm happy, Gus, to see you so clever at pretending, so as to please the heart of this old woman," and once again she didn't look at me as I grimaced with the pain. "I could tell that my future daughter-in-law and her mother would not be able to make batik. So it had to be me who carried out this task. When I was a child, Gus, a woman who couldn't make batik was considered a poor woman indeed."

"Mother's batik is so fine. It must have taken you at least a month to finish?"

"Two months, two batiks. Specially made to be worn on this day. If you both throw them out after today, that is up to you."

"I will save them until I die, Mother."

"How clever you've become in pleasing me. Those are the words of a devoted son. These chains of flowers have also been made by your mother. This keris, this curved ceremonial sword, was left to you by your grandfather. It is hundreds of years old, from before the time of Mataram, from before Pajang. From the time of Majapahit, Gus."

"From where does Mother know this?"

"Hush! Don't be silly, Gus. Don't you remember hearing the family tree explained in your grandfather's house? You never listened to him. That is your fault. Maybe you only value what the Dutch say. This keris has been used by all your ancestors except your father. This keris was prepared for you, Gus. Ah, how must I speak to you? Truly, Mother no longer knows, Gus. Excuse this old woman who knows nothing, Gus."

"Mother!"

"There is no Dutchman who can make a keris, Gus. None are or ever will be able to make one. Open it and you'll see the thumbprints of the craftsman sage who made it."

At that time I was putting on my batik kain, so I said:

"Sorry, Mother, could Mother pull the keris out of its scabbard for me so I can see it?"

"Hush! You're indeed no longer of Java. Do you equate this with a kitchen knife?"

When I saw teardrops on her face I quickly tied up my kain and knelt down before her:

"Forgive me, Mother, it was not my intent to hurt Mother. Forgive me, Mother, a thousand pardons, Mother."

Mother turned away and wiped away the tears from her face.

"Don't go too far, Gus, don't go too far with your non-Javaneseness. Since when has a woman been allowed to pull out a keris from its scabbard? A keris is only for a man. That which is for a woman is not called a keris. Don't be so disrespectful. You too could not make the likes of this. Respect those who can do more than you. Later you can look in the mirror. When you have slipped the keris onto your waist, you will change. You will look more like your ancestors. You will be closer to your origins."

And mother talked and talked. And finally my adornment was finished.

"Sit down there on the floor. Bow down your head," and I knew what would follow on such an occasion as this: the advice before the marriage ceremony. It could be no other way. The advice was beginning. "You are a descendant of the knights of Java . . . the founders and destroyers of kingdoms. . . . You yourself have the blood of a knight. You are a knight. What are the attributes of a knight of Java?"

"I don't know, Mother."

"Hush! You who believe only in everything that is Dutch. The five attributes of the Javanese knight are: house, woman, horse, bird, and keris. Can you remember that?"

"Of course I can, Mother."

"Do you know the meanings of the words?"

"Yes, Mother."

"And do you know what they symbolize?"

"No, Mother."

"Child who doesn't know his own origins, you! Listen, and pass it on to your children one day."

"Yes, Mother."

"First a house, Gus. Without a house a person can never be a knight. He can only be a tramp. A house, Gus, is where a knight departs from, and the place to which he returns. A house is not just an address, Gus, it is a place trusted by he who lives there. Are you bored?"

"I'm listening."

She pulled my ear:

"You have never listened to your parents."

"I'm listening, Mother, truly."

"Secondly, a woman, Gus—without a woman, a knight goes against his nature as a man. Woman is the symbol of life, and the bringer of life, of fertility, prosperity, of well-being. She is not just a wife to a husband. Woman is the center around which circles and from which comes the giving of life, and life itself. This is how you should look upon this old mother of yours, and what should guide you in bringing up your daughters."

"Yes, Mother."

"The Dutch know none of this, Gus. But you must know, because you're Javanese."

"Yes, Mother, they know none of this."

"Third, Gus, a horse. The horse will carry you on your journeys: after learning, knowledge, ability, skills, expertise, and finally—advancement. Without a horse, your strides will not be long, your vision will be short."

I nodded in agreement, understanding too that this was wisdom that had been born out of centuries of experience. Only I didn't really know for sure whose wisdom it was, the ancestors' or Mother's own.

"The fourth, the bird, is a symbol of beauty, of distraction, of everything that has no connection with simple physical survival, of only the satisfaction of one's soul. Without this, people are only lumps of soulless stone. And the fifth, the keris, Gus, the keris is the symbol of vigilance, of preparedness, of courage, the weapon with which to defend the other four. Without the keris, the others will vanish. They will be vulnerable to any attack. You, H.B.S. graduate, your teachers have never taught you any of this? Those Dutchmen? Now you know all that a knight need know. Never be without even one of these things. Do not scoff at the knightly attributes. Each of them is a sign of yourself. You must listen to your ancestors. If you can't obey in the other things, then at least complete the attainment of these five. You hear, Gus?"

"Yes, Mother."

"Now meditate. Ask for the blessings and forgiveness of

your ancestors, that they may guard you from the oppression, slander, and malice of others."

I remained seated on the floor, head bowed.

"Not like that. Sit properly, cross-legged. Your arms relaxed and placed on your lap. Be a good Javanese, even if only for a moment and just this once. Bow down your head more deeply, Gus."

I had carried out all her orders and her wishes. And indeed I did seek forgiveness from those unknown ancestors of mine, ancestors whom I could not even imagine. Instead, just for a moment, Fatso flitted through my mind.

Mother knelt before me, placing a necklace of jasmine flowers around my neck. She was sobbing. Then she placed a small chain of flowers in each of my hands. With her hands, and without speaking, she moved each of my fingers into gripping the chains. She kissed my forehead under the curved edge of my batik blangkon, the sign of Javanese nobility. And her sobbing became worse. I felt her tears drop onto my cheeks. And all of a sudden I too began to cry.

And those images of my ancestors, which had not yet had the chance to even take on faces, faded away, replaced by churning emotions inside my breast, squeezing the tears out in an ever more abundantly flowing stream.

"Bless this child, the child of your blood, my ancestors, your most-favored child. Protect him from disasters, from oppression, slander, and malice, because he is my beloved child. I gave birth to him in suffering, almost dying . . ."

"Mother!" I flung my body to the ground and embraced her knees.

". . . I have lived until today only so that I may witness this event. This is the child of my own blood. Bring him closer to greatness and triumph."

I felt Mother's hands on my back. And Mother ceased her sobbing. She corrected the way I was sitting, the position of the jasmine necklace and the chains of flowers gripped in my hands. With the corner of her kabaya she wiped away my tears. She put right the way I held up my chin, it being too high.

"Meditate, Gus, meditate by yourself, without my help."

Guests started to arrive and fill up the front room, inside living room, and the pavilion set up outside. My own heart was still occupied with the deep impressions left by Mother and the ceremony she had carried out on the eve of my ascent to the wedding throne. I'd never seen a bridegroom undergo such a ceremony. It was probably Mother's own improvisation. Maybe it was a special ceremony for a child who was a renegade in the eyes of his family, but not in the eyes of his mother.

Kommer, who received a special invitation, arrived five minutes before seven o'clock. He strode confidently up to me, put out his hand, and shook mine warmly, then he greeted Annelies, returned to me, and said:

"With this marriage, Mr. Minke, the dirty mouths outside will be silenced. But not only that. You have finished what you started. And the future? We will continue working together, yes?"

"Of course, Mr. Kommer, gladly. We can be good allies. And thank you for your congratulations."

He was a very friendly Indo. Only the shape of his head and the pointedness of his nose manifested his European heritage. The rest was Native, including perhaps his psyche. He was much older than I, maybe ten or fifteen years older. His movements were agile. And from his face you could tell he was not used to staying inside his house.

Jean Marais and May, Telinga and his wife arrived in a hired cart. Magda Peters and my other school friends arrived in the same manner. Mr. Maarten Nijman and his wife came in their own carriage.

The school director and the other teachers did not come. They sent a letter of congratulations with Magda Peters.

One minute before seven o'clock a telegram arrived from Miriam, Sarah, and Herbert de la Croix. And once again I was amazed: How did they know about the marriage?

As I predicted, Robert Suurhof was nowhere to be seen. His absence became the subject of lively conversations amongst my H.B.S. friends.

And Jan Dapperste, always tired and still fed up with his name, was busy running around like a propeller, carrying out his duties as a volunteer waiter.

There were quite a few of my own friends. There was no other Native except for Jan Dapperste.

Mama's customers flocked in. The recent trial, where Mama

emerged as the star of the court had, perhaps, turned out to be an attractive and effective advertisement for the business.

In his usual smooth manner Dr. Martinet carried out the duties of master of ceremonies. At eight o'clock sharp he began very fluently to deliver his speech. He started off by telling of the love between Annelies and me and the great storm with which we had had to contend. Never had he heard of such a storm in any tale of love between two people—a tale well worth being made into a book. (And it was indeed because of that speech that I put together my experiences so that they became this document. . . .)

"This tale is unique," he continued his speech. "It could never be repeated."

One moment that fluent doctor had all his listeners mesmerized into silence, then the next moment he had them all laughing. He emphasized everything important with a movement of his hands. It was a pity he didn't speak Malay—many people couldn't understand Dutch.

Finished with the beautiful telling of the tale of our love, he turned to another and unexpected topic:

"Let us now all look at the portrait hanging above the throne upon which sit the happy newlyweds."

With no less beautiful a movement of his hands, he guided the eyes of all those present to Mama's portrait up above the two of us.

"That painting," he explained, "is a portrait of a Native woman who indeed is extraordinary for her times. Nyai Ontosoroh, a very clever woman, mother of the bride and mother-in-law of Mr. Minke. She is brilliant individual. She is a ship's captain who will never allow her ship to be damaged, let alone sunk. It is through her captaincy alone that this happy occasion is able to take place, the uniting of the gloriousness of a woman with the great skill and ability of a young author. Through this captaincy, two pairs of hands will now proceed forth into each other's clasp for the rest of their lives, as this couple begins what will surely be an equally glorious life in the future.

"And you all know who painted this wonderful portrait hanging here? A painter of great talent! Not just any painter. If we examine the painting carefully we can see that the painter truly understands his subject's spirit. He has brought out the

315

greatness in her. I think that these words of mine are not mistaken. Is it not so, Mr. Jean Marais? Yes, the painter hails from France, a country with a great artistic tradition. Mr. Marais, please stand up."

I saw Telinga help Jean Marais stand up, and everyone started cheering wildly. The Frenchman went red with embarrassment and quickly sat down again.

The doctor's short speech greatly soothed us. And it seemed as if he was launching a propaganda campaign for Mama, and Jean Marais as well.

From where I was sitting, I could also see Darsam, dressed all in black, standing away in the distance. His mustache was lush and shining, and curled up at the ends. His eyes were wandering about everywhere. I couldn't see any machete. But I was sure that there were daggers slipped down under his shirt.

Nyai Ontosoroh, my mother-in-law, sat behind the screen at the back of the wedding throne and never stopped crying. Mother stood beside her daughter-in-law and, without once stopping, kept Annelies cool with a peacock-feather fan.

Behind the screen also, Mrs. Telinga was busy looking after the other women guests.

The pile of presents under our feet grew higher and higher. Who knew from where they all came? Wreaths of flowers were lined up in rows on either side of us. As time went on the rows grew longer.

At nine o'clock the start of the party for the village people could be heard as the East Javanese gamelan bronze orchestra rang out: tayub dance! Every now and then cheering and shouting could be heard. Darsam's men had been ordered to keep watch to make sure no rioting or fighting broke out. And there was palm wine, and it flowed all night.

At nine-thirty people began to make their way home. Dr. Martinet had to leave first because of a call from a patient. About six seconds later, a youth arrived, dressed all in black. His hair shone. A fancy handkerchief adorned his top pocket. A gold watch-chain indicated the presence of a gold watch in his pocket. He strode confidently, dashingly, among all the people preparing to leave. He made his way straight to where the two of us were seated. There could be no mistake: It was Robert Suurhof.

With great politeness, he held out his hand to me and offered his congratulations. Then to Annelies:

"Forgive me for being somewhat late, Mrs. Minke." He bowed down even more politely.

"We're happy you were able to come, Rob," I said.

"Forgive all the things that have happened in the past, Minke," he said, without relaxing his politeness, as if he were a new acquaintance. "Allow me to present a little something to your wife, in celebration of this occasion."

Without waiting for an answer, he took out a gold ring with a very, very big diamond. He took my wife's hand and placed the ring on her finger. He twisted the ring around so the diamond was hidden in her palm. Then he bowed down to her hand, just like in the novels about the Middle Ages. As far as I was concerned he kissed her hand for far too long. Then he faced me.

"I will not go back on a promise, Minke; I admire and respect you greatly, much more than I have ever done before," and he handed me a little box tied in pink ribbon. "This is a little present from me on your wedding day. May you both live happily forever."

"Thank you, Rob, for your kindness and concern."

"I would also like to take this opportunity to take my leave of you. I'm sailing for Europe to study law."

"Safe sailing, happy studying, and success to you."

He walked away in that confident, dashing manner of his and joined the others about to depart for home.

Magda Peters, glassy-eyed, came to excuse herself. She held my hands tightly.

"I would have liked very much to follow your development over the next three years. No matter. If you ever come to Europe—don't forget my address." She walked quickly away from us.

Mr. Telinga and wife, Jean Marais and child did not go home. They stayed the night. Jan Dapperste also. Jan was busy carting the presents to the newlyweds' room upstairs and making a list of the names and addresses of all those who gave something.

Amongst the pile of presents were some things from Miriam, Sarah, and Herbert de la Croix. No one knew who had brought them. There was a little note slipped between them, in Miriam's writing, saying:

Were you embarrassed to invite us? Or perhaps we wouldn't have fitted in properly, my friend? We wanted to stand on either side of that angel so famed for her beauty. What can be done? All we can do now is offer out congratulations, and don't forget to keep up our correspondence. Best wishes, and our regards and all our compliments to your wife.

In Sarah's parcel there was a special letter:

I am leaving for Europe, Minke. I'm fortunate to be able to pass on my congratulations on your wedding day. Adieu! Until we meet again in Europe!

Miss Magda Peters's present included several books and a brochure without the name of the author or of the publisher, and without any year of publication either. Written inside the brochure were the following words:

For a newlywed like you, Minke, the most appropriate things to give are special books not everyone can own, and I have chosen those you will like best. By the time you read this note, I will have already arrived home, and will be too busy to reflect on the happiness of a favorite pupil. May you obtain happiness as you build yourselves a brilliant life together. If at some time you happen to think of your unworthy but sincere teacher, Minke, remember: There has been in this world someone proud to have had a student who has followed in the footsteps of the great humanist Multatuli. But now, Minke, some of the students' parents have succeeded in having me sacked from my teaching job. I have been advised to leave the Indies before I'm actually expelled. I'm leaving tomorrow on an English ship. Good-bye.

"Read this yourself, Jan," I said to Dapperste. "Our teacher." "What is it, Mas?"

"The rumors have turned out to be true. The government is getting rid of Magda Peters, though in an indirect way. It moves me greatly, Ann. Even facing such great troubles, she still made time to come and see us!"

"Expelled from the Indies?" Jan whispered after reading the note.

"Yes, and you don't want to leave Java. Will you do something for us, Jan?"

"Of course, Mas, gladly."

"Will you see Miss Magda Peters off at the harbor, on behalf of both of us, and Mama and you yourself? And for Mother too? Such a kind person should not be allowed to leave without anyone to see her off; it must not happen."

One small, long parcel turned out to contain a beautiful pen with a gold nib. There was a card on which someone had drawn his own picture. Then he had printed the following words:

Greetings and best wishes to the doves Minke and Annelies Mellema, with the hope that you will forgive and forget a person you do not know except by the name: Fatso.

The present fell to the floor.

"Mas!" exclaimed Annelies.

Jan Dapperste picked it up.

"This one is for you, Jan," I said. I put the card into my pocket. I still had to decide whether I would destroy it or keep it for the trial that might resume later.

It was after one o'clock. Jan Dapperste had finished his work. After saying good night, the finale to his evening, he left the room.

I went up to Annelies.

"Now you are my wife, Ann."

"And you are my husband, Mas."

There was a knock on the door. I jumped up and opened it. Mama entered. Her eyes were swollen, as she had had her fill of shedding tears. She approached us and couldn't speak. We understood her intent: to pass on her final words of advice.

"Mama, the two of us would like to thank you for everything that you have bestowed upon us so freely, everything that you

319

have done for us, felt for us, and thought about for us. We will always remember and never forget all this."

She nodded, and went out again.

Annelies came up near me under the gas lamp. She put out her two hands. She didn't want, however, either to be embraced or to embrace.

"This ring, take it off."

I took off that suspicion-arousing ring with its even more suspicious manner of placement.

"You didn't like getting it?"

"I have never answered any of his letters."

In one flash I at last understood how Robert had felt all this time. He too loved Annelies. I studied the ring carefully. It was twenty-two carat gold, with a diamond. It wasn't clear whether the diamond was real or only imitation. It was too big to be a diamond; it was impossible that Suurhof had enough money to give away a present like this. I knew that his pocket money had never been more than twenty-five cents a month. And I knew his parents too—they couldn't remotely be classified as well off. Yes, even his mother had never been seen wearing a ring. And why didn't the ring come with its own box?

So I put it into my pocket.

"Give it back, Mas."

"Yes, I'll return it."

Night marched on. Suurhof and Fatso kept on harassing my thoughts.

19

Science was giving birth to more and more miracles. The legends of my ancestors were being put to shame. No longer was it necessary to meditate in the mountains for years in order to be able to speak to somebody across the seas. The Germans had laid a cable reaching from England to India! And these cables were multiplying and spreading all over the face of the earth. The whole world could now observe the behavior of any person. And people could now observe the behavior of the whole world.

But mankind and its problems remained as they have always been. And no more so than in matters of love.

Take that box that I had in my pocket—a cardboard box lined with black linen. Only two people knew its contents: Robert Suurhof and myself. Neither riches nor money, not diamonds, and no magical charm either. Only a letter from one human being, unsuccessful in gaining a love, to another who had won that love. What can be done! Even in the modern world, how to triumph in love could not be taught in schools.

"Minke, my friend," he wrote in large handwriting, though it was clear his pen shook as he wrote.

He asked forgiveness, expressed his great remorse that he had carried out injustices, had been dishonest, and even acted with malicious jealousy. It's strange, he wrote; the cause of all these acts was not evil but rather a pure and hopeful love for Miss Annelies Mellema. He told how he had seen Annelies five times but had never had the chance to speak to her—even the chance to say hello had almost eluded him. He admitted that he had fallen in love and could not then accept reality. He endured great pain on seeing one Minke so easily gain entrance to Annelies's home and heart. It was not that he had given up hope—he claimed that such surrender was alien to him. He still kept up his hopes. Using all kinds of methods, he had sent her several letters. Not one had ever been answered. He was unable to forget her.

All is over for me now. For you it is just the beginning. I admit I still feel unwilling to give up those hopes. There is no other way to ensure I forget than to leave the Indies. Yes, Minke, I must learn to forget. Even though this is so, don't let the mistakes I have made in the past ruin our relations. . . .

Twenty days after our marriage, a letter arrived from Colombo. Miss Magda Peters reported she had set sail on the same ship as Robert Suurhof. He was working as a sailor and seemed to be ashamed of the fact. Miss advised him that such shame was inappropriate: Such a job was not a humiliation for an H.B.S. student, especially as he had such strong intentions to continue his schooling.

At the same time a letter arrived from Sarah, telling of Singapore and all its wonders: its roadways, clean and wide; busy, yet without dust; and the ships in such great numbers it was as if the harbor did not have enough room. There were far more ships here, she wrote, than I have ever seen in Amsterdam. Even more, she wrote, than in Rotterdam.

On the other hand, a letter from Assistant Resident B——told how his request to the Netherlands Indies government asking that the government help me continue my schooling in the Netherlands had been refused, even though my grades were quite high enough. The main thing required by the government was high moral character. And I didn't fulfill that, he wrote.

That, too, was a fruit of science's progress. Even my moral character had now been given a final unchallengeable brand. First of all by the school. Then by the reports on the course of the trial. I did not, of course, hope for much from other people, yet that brand—so final—still hurt me greatly. I had never done harm to anyone else. I had never harmed another's reputation. I had never done away with other people's goods. I had never dealt in contraband. How was I to defend myself from such arbitrary judgments? Perhaps only Jean Marais taught the truth on this matter: People must be just and fair, starting with what they thought. It turned out that the Europeans themselves, and not just any Europeans either, were the ones who were unjust.

And the modern world had also, perhaps, taken news of me to Europe via those German-made sea cables.

Three months passed. My daily work consisted only of writing in the office and keeping Mama company. Sometimes I also helped her.

Jan Dapperste had received his letter of determination from the governor-general through the resident of Surabaya. His name was now Panji Darman. He was then freed from that hated name of Dapperste. Slowly, gradually, his character also changed—just as he had hoped. He was high-spirited, liked to work, and was generous and open. At first he helped Mama with the office work but then he was moved across to Mr. van Doornenbosch's office to help in running the spice business.

Another month passed. Mother visited us twice.

Five months passed. Sarah de la Croix sent letters to me twice. Miriam also reported that she was going to return to Europe, following after her elder sister. Mr. Herbert de la Croix would be staying on alone in that big, silent residency building and so she asked that I write to him more often.

Six months passed. And then happened what indeed inevitably had to happen: Annelies was summoned (along with Nyai) to appear before the (white) court. Who wouldn't have been startled? The court again. Now it was Annelies who received the main summons.

They left together. I stayed to carry on Mama's work. I wasn't able to finish all that much. I wrote some replies to the military

323

barracks and the harbor master's office as well as the ships' chandlers. I noted down some new orders and some changes of customers' addresses. But the most difficult task was getting rid of the ex-Indies Army soldiers who kept pestering me and who wanted to court Mama.

I once saw Mama turn away some of them four times. The soliders back from Aceh seemed to have made Nyai a major topic of conversation among themselves, and then they would, like soldiers of fortune, set out to catch the rich Mellema widow.

One of these men, an Indo, claiming he was a former *Vanndrig*—a junior lieutenant—even approached me. He had been awarded the bronze medal, he said, and had received ten hectares of good agricultural land near Malang as a part of his pension, and he wanted to become acquainted with Mama. Who knows, he said, perhaps later the two of them could go into partnership. At the end of the meeting, this man who claimed to be a former Vanndrig asked my help: Would I pass on all his words to Nyai? If I was successful, he said, he promised to give me whatever I asked for. This was also a part of my work.

He went away—forgetting to tell me his name.

Otherwise I wrote for *S.N. v/d D.*

They had been gone for more than three hours. I became more and more anxious. I stopped writing. Each time a milk cart came in, I went outside to have a look.

Four hours passed. The carriage I had been waiting for arrived at last:

"Minke, quickly!"

I ran to meet her on the front steps. Mama got out of the carriage first. Her face was scarlet. She put out her hand to Annelies, who was still inside. And then out came my wife, ghostly pale and her face bathed in tears, mute. As soon as she alighted she fell into my arms and embraced me.

"Take her in!" Mama ordered me, roughly.

She strode quickly inside and went into her office.

"Have you had a fight with Mama?" I asked Annelies.

She shook her head. But no sound came out of her mouth. I went to take her upstairs. Her body was cold.

"Why is Mama angry?" I asked.

She didn't answer. And she refused to be taken upstairs. She asked with her eyes to be put down on the front-parlor settee.

"Are you ill, Ann?" and she shook her head. "What's the matter with you?" and I became anxious that something had seriously upset this fragile doll of mine. She seemed to be in a state of total confusion. "I'll get you something to drink."

I fetched a glass of water for her. She drank it, and it seemed that the tightness in her chest subsided.

"Darsam!" exclaimed Mama from the office.

I ran to fetch the Madurese fighter. I found him at home, in the middle of plucking out the unwanted hairs of his mustache.

"Quick, Darsam, Mama is angry."

He jumped off his chair. The little mirror and the tweezers fell onto the mat. By the time I reached the office he was already inside. Annelies too.

"Why don't you go to sleep, Ann?" Nyai admonished her hurriedly. My wife shook her head. Mama's face was still scarlet.

"What's happened, Mama?"

Darsam saluted Nyai and departed from the office. There must have been a carriage ready, because it was only a second before the sound of wheels could be heard grinding the pebbles on the roadway in front of the office.

Mama paid no attention to my question, went over to the window, shouted outside:

"Hurry! Be careful!" She turned around and went across to Annelies, caressed her hair, and generally tried to humor her. "You don't need to think about it. Let us look after it, Ann, me and your husband." Then to me, "It's come at last, child, Minke, Nyo, what I've worried about all this time. I don't know much about the law. But we must try to fight it with all our strength and all our wealth."

"What's this all about, Mama?"

She pushed several documents into my hands, originals and copies, from the Amsterdam District Court, stamped by the Bureau of the Home Affairs Ministry, the Ministry for Colonies, the Ministry of Justice. On top of the pile was a letter from Engineer Maurits Mellema in South Africa to his mother, Amelia Mellema-Hammers, giving power of attorney to the latter party to make all arrangements in relation to rights of inheritance from the late Mr. Herman Mellema, his father, who had been killed in Surabaya, as he had been informed in a letter from his mother. Then there was a copy of a letter from Maurits Mellema's mother, written on behalf of her son, to the Amsterdam court, asking it to look after the rights of her son over the wealth and property of the late Mr. Herman Mellema.

Furthermore: copies of the Surabaya court and prosecutor's office's correspondence with the Amsterdam court, concerning whether or not there was a marriage certificate for Herman Mellema with Sanikem, whether or not there was a will made up by Herman Mellema before he died, the decisions of the court in relation to the murder carried out by Ah Tjong, a determination relating to the disappearance of Robert Mellema, copies of the certificates of acknowledgment by Herman Mellema of his children, Annelies and Robert Mellema, both of whom were given birth to by Sanikem as registered in the civil registry office. Then again there was correspondence between Nyai's accountant and the Surabaya court all relating to the accountant's refusal to make available any information relating to the assets of Boerderij Buitenzorg without permission from those with the proper authority. Copies of documents from the Tax Office regarding the amount of taxes paid by the company. Copies of documents from the Livestock and Agriculture Office regarding the number of cattle and their condition.

I read each of the letters one by one under the gaze of Mama and Annelies, who seemed to be hoping for some opinion from me. I didn't know anything at all about any of the things mentioned in all those letters. And, indeed, I had never even dreamed that such letters existed in this world. And I never knew that there were people who were actually paid to write them.

Then there was the official document of the Amsterdam District Court. Its contents: an order that its decision on the case was to be executed by the Surabaya District Court. In brief it read:

Based upon the application to the Court by Maurits Mellema, son of the late Herman Mellema, made through his attorney, Mr. Hans Graegg, located in Amsterdam, the Amsterdam District Court, based upon official documents, provided by the Surabaya District Court, whose authenticity cannot be doubted, determines that the entire property and wealth of the late Herman Mellema, because of the absence of legal ties between Herman Mellema and Sanikem, be divided as follows: Maurits Mellema, as the legitimate child, to receive ⅘ of all property; Annelies and Robert Mellema, as legally recognized children, to receive ⅕ each. Because Robert Mellema's whereabouts have been officially declared unknown, both temporarily as well as permanently, his inheritance is to be managed by Mr. Maurits Mellema.

The Amsterdam District Court also appointed Mr. Maurits Mellema guardian over Miss Annelies Mellema, as the latter is still considered to be legally under age, and so therefore her inheritance will also be managed by Mr. Maurits Mellema. Mr. Maurits Mellema, as guardian of Annelies Mellema, through his attorney, Mr. Graegg, authorized another advocate located in Surabaya to bring action against Sanikem alias Nyai Ontosoroh, and Annelies Mellema, in the Court in Surabaya, over the guardianship of Annelies Mellema and her future upbringing in the Netherlands.

I felt as if I was about to faint as I read those official documents with their strange language. I could, however, understand one aspect of their contents very well: They looked upon human beings as no more than items in an inventory.

"Mama didn't say anything to them?"

"Look, Minke, child, Nyo, my attorney was waiting for us there when we arrived. He was the one who arranged to get all these letters. He, too, was the one who, before the judge, told us of the decision and explained it."

As I listened, the words of Mother came back to me: "The Dutch are very, very powerful but they have never stolen people's wives as did the kings of Java." But now, Mother? It is none other than your own daughter-in-law they are threatening to steal, to steal a child from her mother, a wife from her husband; and they want, too, to steal the fruits of Mama's hard work and everything she has strived to achieve over the last twenty years without ever a holiday. And all this was based upon no more than beautiful documents written by expert scribes and clerks with their indelible black ink that soaked halfway through the thickness of the paper.

"It looks like we'll have to get help from a lawyer, Mama?"

"Mr. D—. will soon be here, I think."

That strange name had already made itself familiar during my recent complex and multifarious problems.

"Mr. D—. L—."

For quite a long time I had tried to learn his name by heart and to write it. I had never met the man himself. Mama often went to him for legal advice. My image of him was of a big and fat man, like Herman Mellema, with thick blond hair all over his body. His name reminded me more of some kind of spirit. He must surely be a brilliant lawyer.

"Didn't Mama protest against the decision?"

"Protest? I did more than that—I completely rejected the decision. I know them, those Europeans, cold, hard like a wall. Their words are expensive. She is my child, I said. It is only I who have any rights over her. It was I who gave birth to her, who have brought her up. The judge only said: The documents show that Annelies Mellema is the acknowledged child of Herman Mellema. Who is her mother, who was it that gave birth to her? I asked. The documents state that her mother is the woman Sanikem alias Nyai Ontosoroh, but. . . . I am Sanikem. Yes, he said, but Sanikem is not Mrs. Mellema. I can bring witnesses, I said, to prove that I gave birth to her. He said: Annelies Mellema is under European law, Nyai is not. Nyai is a Native. Had Miss Annelies Mellema not been legally acknowledged by Mr. Mellema she too would be a Native and this court would have had nothing to do with her. Minke, what could be more humiliating! So I said, I will fight this decision, using whatever attorney is able and willing. That's up to you, he said coldly. Annelies just cried and cried, so that I forgot about everything else."

She took a deep breath.

"You should have come too, Minke. You could have at least defended your wife and your interests, even if the court wasn't actually in session. Yes, even the judge has a wife, and children, too."

I am sure that everyone will know how I felt at that moment: angry, furious, annoyed, but not knowing what I had to do. In such matters I was still a snotty-nosed little boy.

"I said too: My child is already married. She is somebody's wife. He then just smiled, just a shadow of a smile, and answered:

She is not yet married. She is under age. If there has been some-body who carried out a marriage of her to somebody or married her, the marriage is not legal. You hear, Minke? Not legal."

"Mas?"

"They then threaten to charge me as an accessory to rape because I hadn't reported the marriage as illegal."

The office was still, silent. There were no customers about.

The three of us were silenced. Once a brilliant and honest attorney would have been able to appeal successfully against the decision of the Amsterdam District Court. Oh, Amsterdam court! You had never even seen us. How can a court, and a European court too, manned by very educated people, experienced in mat-ters of justice, with the degree of Bachelor of Laws, carry out the law this way, so opposed to our sense of law? Our sense of justice?

"I didn't even get on to talking about the division of the property. Yes, indeed, even though the land was bought in my name, I don't have enough documentation to prove to a European court of law that the company itself is my property. All I tried to do was to defend Annelies. At the time, all I could think of was her. Actually our business is only with Annelies, the judge said. You are a nyai, a Native, you have no business with this court." Mama grimaced savagely.

"In the end," she said later in a soft voice, "the issue is always the same: European against Native, against me. Remember this well: It is Europe that swallows up Natives while torturing us sadistically . . . Eu-r-ope . . . only their skin is white," she swore. "Their hearts are full of nothing but hate."

"And the attorney, he's a European too, Mama?"

"Just a slave to money. The more money you give him, the more honest he is with you. That's Europe."

I shuddered. Years and years of schooling were overturned with just the three short sentences of a nyai.

Annelies had fallen asleep, exhausted by all the emotional tension; she lay with her head on top of the table. I went up to her and woke her up.

"Let's go upstairs, Ann."

She refused to move—and sat up straight again in her chair.

"Get some sleep, Ann. Let us look after you," entreated Mama, and Annelies complied.

I took her upstairs, put her to bed, and began to humor her: "Mama and I will work hard, Ann."

She just nodded, and I knew with all my heart I had lied to her—I knew nothing about all the ins and outs of the law. How then could I work hard at it?

"You stay here, yes, Ann?"

She nodded again. But I couldn't bring myself to leave her in this condition—like a fish that was already in the frying pan. How moving was the fate of this fragile doll, my wife. It looked as if she had lost the will to do anything at all.

"I'll call Dr. Martinet, yes, Ann?"

She nodded.

I went downstairs and gave orders that someone fetch the family doctor. I saw Marjuki racing his buggy off towards Surabaya.

In the office, Mama was with a European man who was small-bodied, like one's little finger, perhaps only up to my shoulder in height, thin and flat. His head was slippery bald, and his eyes just a little slanted. He wore horn-rimmed glasses. Mama was watching him reading the documents regarding Annelies from the Amsterdam District Court. So that was Mr. D—. L—. It was clear he was no spirit. And he had been Mama's lawyer all this time.

I was amazed Mama wanted to deal with him. Just a while ago, before the judge, he hadn't said or done anything, had he? I watched them both. Mama was no longer so red-faced. Her movements were calmer now.

"Minke, this is Mr. D—. L—. . . ." and we shook hands. "This is Minke, the husband of my daughter, my son-in-law."

"Ah yes, I've heard a lot about you. Could I just finish studying these documents first?" and without waiting for an answer, he resumed his work.

A person no bigger than one's little finger, with a face full of craters from the explosions of so many pimples—just how far would he be able to go in confronting the arbitrariness and might and coldness of European law and justice? And as a European, on whose side would he ultimately stand?

And he studied the documents one by one, turning them over and reading them again.

Mama rushed about finishing her own work, then even served

him drinks. And the lawyer kept on with his inspection of the documents as if nothing at all was going on around him.

Finally, one hour later, he piled the letters together and put a paperweight on top of them, a black stone. He meditated importantly, wiped his face with his handkerchief, and then gazed across at me, then at Mama, and he didn't say a word.

"So what about it, Mr. L—.?" asked Mama. "Oh, I'm sorry, I'm still not sure how to pronounce your name properly."

He smiled—just briefly—which turned out to be because of missing teeth.

"Oh, that's all right, that's just my name for signatures, Nyai. I don't mind if people are unable to pronounce it. It doesn't even worry me if they don't even try."

"You can still make jokes while we're suffering a situation such as this, Mr. L—! We're already half crazy with all this!"

"That is the way it is, Nyai, when the matter is a legal one. There is no point in changing one's feelings or countenance. The result is just the same whether people laugh, jump up and down, or cry and wail. It is always she who determines things in the end: the law."

"So we will be defeated in this matter?"

"It's better we don't talk about defeat, Nyai," said the attorney and his hands began to finger the documents once again. "We haven't begun trying yet. What I meant was that I hope Nyai will be as cold and calm as the law itself. Feelings have no influence over any of this. All anger and disappointment is in vain. Are you listening too, sir?" Suddenly he turned and faced me. "You understand Dutch?"

"I'm listening, sir."

"This concerns the fate of your wife and your marriage. The other side is in the stronger position. We will try if you and Nyai still want to fight the decision; at the very least we will get its execution postponed."

I understood at that moment: we would be defeated and our only duty now was to fight back, to defend our rights, until we were unable to fight back any longer—like the Acehnese in their fight against the Dutch according to Jean Marais's story. Mama also bowed her head. She more than just understood. She was going to lose everything: her child, her business, all the fruits of her efforts, and her personal property.

"Yes, Minke, child, Nyo, we will fight back," whispered Mama. And all of a sudden she looked old, and walked dispiritedly upstairs to check on her daughter.

Mr. D—. L—., LL.B, submerged himself once again in his study of the documents. My suspicion of this little finger-sized lawyer suddenly swelled to become so great that I stayed and watched his hands closely to make sure he did not secrete any of those documents.

Another hour passed. Mama came down again and entered the office. She sat beside me, across from the jurist.

"Do you still need to study them, Mr. L—.?" she asked in her old voice—one with character.

The man lifted up his head, held back a grin, and said:

"We can try, Nyai."

"You do not believe we can win."

"We can try." He started his reading again.

Mama took the letters from him.

"Your fee will be sent to your office. Good afternoon."

Mr. D—. L—. stood up, nodded to us both, and was then escorted home by Darsam.

"Minke, we will fight them. Do you have the courage, child, Nyo?"

"We will fight, Mama, together."

"Even if we don't have a lawyer, we will be the first Natives to oppose the European court, child, Nyo. Isn't that also an honor?"

I had no idea of how I was to fight back, what I had to fight, who and how. I did not know either what should be my tools or through what mechanism we should endeavor. But: We'll fight!

"Fight, Mama, fight. We will fight back."

"If you could only get Annelies to get up and fight too, she wouldn't always be falling into illness and incapacity like this. She would become the best of all life companions for a husband such as you."

While waiting upon Annelies, I let loose my thoughts and allowed myself to concentrate on everything that had happened and was happening.

Engineer Maurits Mellema and his mother did indeed have reason to seek revenge on Herman Mellema. But what was happening now? Their revenge did not make them spurn his inheritance; rather they wanted to make sure they got every cent. In fact, they too had wanted Annelies's papa to die. In their hearts they too had participated in and approved of Ah Tjong's actions. But they would not be punished. The life of the soul and the psyche are not mentioned in official letters.

Yes, this was nothing more than a case of the white race swallowing up Natives, swallowing Mama, Annelies, and me. Perhaps this was what was called a colonial case, if Magda Peters's explanation of things was right—a case of swallowing up a conquered Native people.

Suddenly I was reminded of the liberals who, according to my teacher, wanted to lessen the sufferings of the Natives. That, too, was what the S.D.A.P., the Dutch Social Democratic Workers Party, wanted. Ah, my good teacher. I regret now that I didn't go to see you off. If you were still in Surabaya, you would surely hold out your hand in assistance. At the very least you would offer us some guidance to help us. And you would do so gladly.

And as I thought about Magda Peters, there arose suspicions that were perhaps too fantastic: She had been forced to leave the Indies to make the execution of the Amsterdam District Court's decision easier! Perhaps you weren't really exiled but just gotten out of the way for the coming case. My suspicions then took on an even clearer form: Everything had been arranged beforehand by that satanic alliance between Maurits-Amelia and the Amsterdam District Court. If it was true that Magda Peters had been gotten out of the way, then it was the school director and the other teachers who knew how close she and I were. If my suspicions were correct, this whole thing, everything, was no more than a prearranged drama whose purpose was the sadistic torment of human beings. So even my graduation as number two in all of the Indies (to be number one was out of the question) was also no more than a play, manufactured to keep the radicals or the S.D.A.P. happy.

Should I have such grandiose suspicions as this? Was my thinking, as an educated person, fair and just? Was I not both too young and too ignorant to have such suspicions? I thought it over and over again. I couldn't avoid always coming back to the same condition: being inclined to accept my suspicions. My dismissal from school, the withdrawal of my dismissal, the closing down of the school discussions, the expelling of Magda Peters, the intervention of the assistant resident of B——, the invitation that was announced by the school director at the graduation party, and his own and the other teachers' absence at our wedding, and being represented only by the letter brought by Magda Peters. . . . No, I was not too ignorant or too young to understand. Each event was linked and entwined together so as to give victory to Maurits Mellema over the Native woman Sanikem, her daughter, and son-in-law, her wealth and property.

"You've got some ideas, Nyo?"

"Mama, this evening, if nothing goes wrong, my first writings on this affair will be published. If it is not greeted by common sense, Mama, we will be defeated. We need time."

"Don't think about defeat, Mr. L—. said, think first about the best way to fight back, the most honorable resistance. Mr. L—. was right, it was only his motives that were wrong. He only wanted bigger fees. That shrunken-up crocodile."

"We'll turn to the best of our European friends, Mama."

"Don't make any mistakes."

That afternoon I sent a cable to Herbert de la Croix, appealing to his conscience regarding our case. Also to Miriam. If no one wanted to listen, then I would know: All that glorified European science and learning was a load of nonsense, empty talk. Empty talk! In the end it would all be nothing more than a tool to rob us of all we loved, all we owned: honor, sweat, rights, even child and wife.

That night Mama and I waited upon Annelies, who had to be drugged once again by Dr. Martinet so that she could sleep. The doctor was moved and saddened by the situation of his patient, her mother, and her husband, who were bound so tightly by man-made fate, a fate manufactured far away in the north.

"I am only a doctor, Nyai. I don't know about the law. I don't know about politics," he said, expressing his disappointment in himself.

He was the second person who had mentioned the word politics.

"It's only proper that I ask your forgiveness because I am unable to do anything to help lighten your sufferings. There are no important people among my close friends, because I have never joined any of the clubs."

And in what a modest manner the doctor presented himself.

"My only friends are those who have needed my help. I don't have anything more than that. I'm sorry."

"But you feel that the way we are being treated is unjust, yes?" asked Mama.

"Not just unjust. Barbaric!"

"That is enough, Doctor, if it has come from a sincere heart."

"Forgive me, there is nothing I am able to do."

He left us with such a pained face. At the door he spoke in a sighing voice:

"I used to think that the only real difficulty in the world was paying one's taxes. I never knew that under these heavens there could exist difficulties such as these."

He disappeared into the darkness, escorted by Darsam.

Five hours had passed since I sent off the telegrams to the de la Croixs. Five hours! And still no answer had arrived. Were Herbert and Miriam de la Croix not at home? Or were they laughing at us Natives?

"Yes, child, Nyo, we must fight back, we must resist. However good and kind any European has been to us, in the end they will be afraid to face up to the risks of resisting European law, their own law, especially if it's only to defend the interests of Natives. We need not be ashamed if we are defeated. We must know why. Look, child, Nyo, we, all the Natives, are unable to hire attorneys. Even if we have money it doesn't mean we are able to do so. The main reason is that we don't have the courage. And more generally still, we haven't learned anything. All their lives the Natives have suffered what we are now suffering. No one raises their voices—dumb like the river stones and mountains, even if cut up and made into no matter what. What a roar there would be if they all spoke out as we will now speak out. Perhaps even the sky itself would be shattered by the din."

Mama had begun to forget her own feelings. She was placing the matter in the context of a more basic problem. She had left behind her own heart and her family, she had now brought in the river stones, the mountains, the chalk and granite rocks that were strewn all over Java, throughout all the Indies; those with mouths but no voices, and yet with hearts within them.

"By fighting back we will not be wholly defeated," and the tone of her voice was pregnant with the knowledge of coming defeat.

335

"They know no shame, Mama."

"Shame is not a concern of European civilization." Mama stared wide-eyed at me as if she were angry with me. "You who have mixed with them all this time, how can you talk like that? You, child, Nyo, as a Native, should and must be ashamed to have such thoughts. Never again mention shame in relation to Europe. All they understand is getting their way. Never forget that, child, Nyo."

"Yes, Mama," I answered, acknowledging her superiority. The truth or otherwise of what she said was, of course, another matter.

"I've never been to school, child, Nyo. I've never been taught to admire Europeans. You could study for years and years, and no matter what you studied, your spirit will be educated to do the same thing: to admire Europeans without limits or end, so that you no longer know who you are and where you are. Even so, those who have been to school are still more fortunate. At the very least you get to know other races who have their own ways of thieving the property of other peoples."

My mother-in-law took a newspaper from the table. Inside, there was an article of mine, and comments on it by the editor.

"Your writings are so gentle, like the writings of a teenage girl waiting for a husband. Have you still not become hard with all your recent experiences, let alone this current one? Uncompromisingly hard? Minke, child, Nyo," she went on in a whisper, as if there were someone else there who was listening in on us, "now you must write in Malay, child. The Malay papers are read by many more people."

"It's a pity, Mama, I can't write in Malay."

"If you're unable at the moment, let someone else translate for you."

And straight away I thought of Kommer.

"Good, Mama," I answered quickly.

"Your marriage is legitimate according to Islamic law. To nullify it is to insult Islamic law, to besmirch the laws honored by the Islamic community. . . . Ah, how I dreamed of a legitimate wedding for myself. Mellema always refused because he already had a wife. Now my child has married legitimately, more honorable than me. And it's not acknowledged."

"I'll work on it now, Mama. Mama should get some sleep."

And she went off to bed. Her strides were still strong and firm like those of an undefeated general.

It was ten minutes past three in the morning. My article was almost finished. Out of the predawn silence came the pounding of a horse's hooves, coming closer and closer, and finally entering our grounds. Not long after Darsam was calling out from below my window.

"Young Master, wake up!"

Below, in the light of an oil lamp held by Darsam, I saw Darsam with an Indo in the uniform of a postman. He saluted, and asked in Malay:

"Tuan Minke? There is a telegram from the assistant resident of B——"

He left happily with a tip of five cents. The pounding of his horse's hooves disappeared in the distance to the accompaniment of the cock's crows.

"Young Master has already done a lot of work. It's already dawn. Get some sleep, Young Master. There will still be other days."

He didn't know a thing about what was happening. But I could sense he was anxious at seeing all the activity that was going on. Ah, Darsam, a thousand such as you, even with two thousand machetes, would be unable to help us. This is not a problem of flesh and steel, Darsam. This was a matter of rights, law, and justice—you cannot protect us with dagger and machete. Suddenly there was a reprimand: You must be fair and just, starting with how you think. Darsam the fighter with his machete, even the mute stones and rocks can help you—if you know and understand them. Never belittle the capabilities of a single person, let alone two.

"Very well, I'll get some sleep now, Darsam."

"Yes, go to sleep, Young Master. A new day will bring new opportunities."

How wise too was this black-clothed man. I went upstairs and read the telegram:

A WELL-KNOWN JURIST WILL ARRIVE FROM SEMARANG THE DAY AFTER TOMORROW. TRUST HIM. MEET HIM AT THE STATION. EXPRESS TRAIN. GREETINGS TO NYAI AND ANNELIES. MIRIAM AND HERBERT.

Mother! Mother! at last my cries have been heard. And you yourself have not even heard what is happening. Sleep deeply, Mother. I will not awaken you. Not now either. And here, your beloved son will not run. He will stay and fight. He is no criminal, Mother. Your beloved daughter-in-law will not be stolen away. She will present to you the grandchildren you long for, so one day you will be able to attend their weddings as Javanese.

An article about the contravening of Islamic law by European law appeared in Dutch in *S.N. v/d D.* Malay versions appeared in the Malay-Dutch press. They all appeared on the same afternoon. Mr. Maarten Nijman himself came around to our house to deliver the complimentary copy.

"You have helped us a lot all this time. Now it is our turn to help you as much as we can," he said. "But there is nothing else we can do to help lighten your own and your family's burdens. All of the editorial staff and the workers at the paper have high regard for your resistance, and express their true and sincere sympathy—so young, like a sparrow harassed by a storm, but yet still fighting back. Another person would have been broken even before the fight started, Mr. Tollenaar."

He borrowed a picture of Annelies to publish.

"If possible, also a picture of you and Nyai."

Mama give him a big picture of my wife in full Javanese dress adorned with diamonds and pearls.

"It's only a pity that we won't be able to publish the picture soon. We'll have to wait almost two months," Nijman explained. "The Indies is still a wilderness. There is no factory here that can copy this picture onto tin: zincography is still not yet known here. We'll get the negative made in Hong Kong. If Hong Kong can't do it because of all the orders from Southeast Asia, then we'll have to send it to Europe to get done. Longer still. If we succeed it will not only mean greater impact, but we will be the first in the Indies to publish from a tin negative."

He talked a lot and asked to be introduced to and meet with Annelies herself. And we refused on the grounds that she was ill.

"Is Miss Annelies with child?" asked Nijman. "Forgive the question. It might seem improper, but it could change the situation. It could nullify Maurits Mellema's decision even if the Amsterdam court's decision still stood."

Annelies pregnant? I had never even thought about it. I couldn't answer. Neither could Mama; instead she looked questioningly at me.

After Nijman left, Kommer arrived, also bringing a complimentary copy of his paper.

"Nyai, Mr. Minke," he said, "your writings will soon be in the villages. We've hired men to read them out to the people. People gather around and listen. Fifteen special copies with the relevant parts underlined in red have been sent off to the leading Islamic scholars. They must also speak out. I'm going to try to get their opinions tonight. You and Nyai will not stand alone. Look upon Kommer here as a friend of the family in its time of trouble."

We set off to Surabaya together. He got off at Gunungsari. I went on to the station to meet the attorney whose name I still didn't know. Kommer, before I left him, shook my hand from outside the buggy. His eyes shone with his enthusiasm for the humanitarian task he was undertaking. Then he waved his hands, and my buggy started off again.

The attorney I met turned out to be middle-aged. He had a calm demeanor and he smiled a lot, and liked to listen, not like Mr. D—. L—. His name was Mr.——. I'll not mention his name here either. He was a famous jurist and a very wealthy man as a result of his practice as a brilliant attorney and advocate, and his name was often mentioned in connection with many big cases.

He stayed at our house. He studied Annelies's file all night, and asked that two scribes be hired to make copies of every document. Panji Darman, formerly Jan Dapperste, and I acted as scribes. But I was sacked in the end because of my bad handwriting and because I made so many mistakes. So Darsam had to go out that night and find a clerk from the D.P.M., who also brought the special ink used for official documents.

Mr.—— (whose name I don't dare mention; and who could tell if he might be unsuccessful in this case and his practice affected) studied it all until morning. The scribes made two copies of each document. At six in the morning they had to leave for their regular jobs, so we had to hire two more people.

At seven o'clock in the morning Mr.—— began to write a long letter, of which the new scribes made several copies. Taking one set of the copies, he headed off to the European court in

Surabaya with Darsam. He arrived back again in the evening and went straight to sleep.

We didn't know what had happened at the court.

The afternoon news, as published by Kommer, reported that the Islamic scholars had gone to the European court at Surabaya to protest the decision of the Amsterdam District Court and its execution by the Surabaya court. They threatened to take the matter to the Islamic Religious Supreme Court in Betawi. And they were removed by the police especially brought in for that necessity.

The commentary, which seemed to have been written by Kommer himself, warned that it would be wise for those in power to act more tactfully in dealing with the Islamic scholars who were held in respect, honored, exalted, and listened to by the followers of Islam in this region. It is dangerous to play with the beliefs of the people, much more dangerous than to make fun of powerless subjects of the realm or rob them of their rightful property and their women and children.

For the second time Kommer emerged as a friend. He was so skillful at speaking for us, for our situation, and for the general conditions of the Natives. His words were so simple and moving, yet confident and full of substance. And not without risk.

S.N. v/d D. published an interview between Nijman and Nyai:

For more than twenty years now I have worked my bones, building, defending, and keeping alive this business, both with and then without the late Herman Mellema. I've looked after this business better than I have my own children. Now it is all being stolen from me. The attitude, illness, and incapacity of the late Mr. Mellema resulted in my losing my first child. Now another Mellema is going to steal my youngest also. Through the use of European law, he is having me torn from all that is mine by right, and all that I love. If that is indeed his deliberate intention towards us, all I can do is to ask: What is the point of having all these schools if they still don't teach what are people's rights and what are not, what is right and what is not?

And he wrote up his conversation with me as follows:

We married of our own accord, and our marriage was approved of by the girl's parents. Our persons are our own property; we are nobody else's property; slavery was abolished by law in 1860, or so, at least, we have been taught in our history lessons. Now with the impending kidnapping of my wife, in accordance with the court's decision, I wish to ask the conscience of Europe: Is that accursed slavery going to be brought back? How can human beings be looked upon purely from the point of view of official documents and without considering their essence as human beings?

Then there was an interview with Dr. Martinet:

I have known this family quite some time now. So I understand the situation of Annelies Mellema's health, both before she married, and afterwards. With a heavy heart I have to say that this girl loves her husband, her mother, and her surroundings very much. She is very, very attached to all three. If indeed the Amsterdam District Court's decision is executed, the life of this girl could be destroyed through the emotional turmoil that will result. Even now Annelies has to be sedated. She has lost all faith in the existence of security, certainty, and legal guarantees. Her spirit has been crammed full of fears and uncertainties. Must I continue to drug her while outside there is the sun, laughter, and joy? Why must this young angel become the plaything of decisions that have no real connection with her life and happiness? As a doctor I cannot accept the responsibility for what might happen if I have to continue drugging her.

The attorney from Semarang, Mr.——, read through everything that was connected with our case. He made notes but never spoke to us. Nor did we bother him with any questions. In the evening he read the papers from other towns. Only after all this did he finally begin to speak about many things.

"We must be resolute, Nyai . . . sir . . ." Then he asked Mama:

"Why, really, didn't Mr. Mellema ever marry Nyai legally?"

Mama answered:

"I didn't understand why Mr. Mellema didn't want to marry either, although I often pressed him to marry me. I only realized what the situation was after the sudden arrival here, at this house, of Maurits Mellema, his son, five years ago. Only then did I understand that Mr. Mellema was still legally bound to the mother of that engineer."

Mr.—— turned and looked at her in amazement.

"So they were never divorced? If that's so then it was impossible for Mr. Mellema to acknowledge his children legally here, because such children are considered bastards and acknowledgment of them is not considered legal. But if that's so, then Nyai's position in this case is much stronger!"

Within Mama and me feelings of hope were awakened once again. Mama was angry: Why had not Mr. D——. L——. thought of that point? But some days later Mr.—— reported to us that such a defense would not help us either.

Mr.—— said: "After sending off a telegram to Holland checking things out, it has been shown that Mrs. Amelia Mellema-Hammers, after her husband had left her for five years without giving any address, did apply for a divorce in the Dutch courts on the grounds of desertion. After efforts to find Mr. Mellema were unsuccessful, in 1879 the divorce was granted. So the marital ties between them were already nullified when your child, Robert, was born." Then he asked: "Did Mr. Mellema know of this divorce?"

"I don't think so," answered Mama. Mama thought for a moment, then exploded: "If this is all true then Maurits Mellema lied to his father when they met that time five years ago! He challenged his father to institute divorce proceedings against Mrs. Mellema on the grounds that she had been unfaithful. He destroyed his father's spirit with that conversation."

With fury in her eyes Mama sat silently, not saying a single word, but I saw her hands shaking because of the overflowing of her emotions. Our picture of Maurits Mellema was only getting worse. It seems he deliberately set out to destroy his father's spirit and so speed up his death. And all for money.

The next morning Mr.—— returned to Semarang.

We were left without the support of an attorney, without any direct means of fighting the court's decision.

"All right, Mama, only the pen is left," and so I wrote, calling out, speechifying, complaining, roaring, swearing, crying out in pain, agitating.

Kommer translated them all and gave them out to those who were prepared to publish them.

And it was not without results.

The Religious Supreme Court in Surabaya issued a statement: Our marriage was legitimate and could not be disturbed or nullified. On the other hand, some of the colonial papers started flinging insults, curses, and slights at us. Nijman's and Kommer's papers were busy summarizing all the various statements.

While Annelies, my wife, that fragile doll of mine, was lying like a corpse on her bed. Surabaya was in a fever over her, Nyai's, and my troubles. Kommer kept on fighting too. His paper was being read, and also read out aloud in the villages, and big crowds of people stopped to listen everywhere. Without going by way of eyes, without going by way of ears and mouth, the news had spread and had become a matter of wide public controversy.

Finally Darsam also found out what was happening without ever having to ask us. He was busy reading the Malay papers with the help of his children.

Once again Annelies and Nyai received a summons from the court. It was impossible for Annelies to go. Only Mama and I went, unaccompanied by an attorney. Dr. Martinet waited upon my wife.

The judge immediately asked where Annelies Mellema was.

"Ill. In the care of Dr. Martinet."

"Have you brought a letter from the doctor?"

I was startled to hear Nyai answer coarsely:

"Has the court already decided that my mouth cannot be trusted?"

"Good," answered the judge, red-faced. "Nyai should be more polite."

"Should someone about to lose everything show politeness in the face of her loss? Just tell us what you want."

The judge deliberately avoided a clash with the Native woman. He gave in.

"Good. In my hand is the Surabaya court's decision regarding Miss Annelies Mellema, the acknowledged child of the late Mr.

Herman Mellema. In accordance with this decision, Miss Annelies Mellema is to be transported from Surabaya by ship in five days' time."

"She's ill," answered Mama.

"There are good doctors on board."

"I refuse to let her go. I'm her husband."

"We have no business with anyone who claims or who doesn't claim to be her husband. She is still unmarried, without a husband."

There was no way to get this devil to be reasonable. He took out his pocket watch, rose from his chair, and left us.

The two of us left the building in complete anger. I asked Mama to go home first. I got in touch with Kommer and Nijman to tell them the news, and even helped in getting the report ready, right up to setting the headlines.

That afternoon the news was published.

I found Dr. Martinet waiting upon Annelies and Mama. The two of them sat silently, heads bowed. Neither of them seemed to want to talk at all.

The next morning something amazing occurred.

The Surabaya court's decision had angered and infuriated many people and groups. A crowd of Madurese, armed with machetes and large sickles, had surrounded our house, and were attacking any Europeans or state employees who tried to enter our compound.

The traffic felt it had to stop to watch what was happening.

A Madurese, wearing all-black clothes, walked back and forth with his shirt open, showing his chest, as if it was being deliberately readied to fight anyone and face any risk. The tip of his headband with its long tail fell over his shoulder.

From Annelies's window they could be heard continuously cursing and condemning the white court's decision as the act of infidels, as sinful, damned in this world and the next. From early in the morning until eleven o'clock they controlled our compound.

All the activities of the business stopped. The workers dispersed in fear and went home to their villages.

Two companies of police arrived, escorted by government carriages. The ringing of their copper bells could be heard from afar. Not paying any heed to the Madurese, the carriages came straight into our grounds. We could see from our room some of the Madurese swinging their great sickles against the legs of the horses. Two carriages went out of control and into the garden, splashing into the swan pond. Out of the carriages that succeeded in entering the yard jumped uniformed men with carbines who tried to disperse the Madurese. Those under attack did not want to leave. A fight took place.

From where I was, I saw two policemen felled, bathed in blood. The uniformed men finally didn't know what else to do and fired off their rifles into the air.

Here and there could be seen a Madurese laid out, also covered in blood.

The police commandant, a Pure-Blood, swore at his men for firing their rifles. A stone flew through the air and struck his temple. He swayed about, fell, and did not rise again. A black Dutchman, an Ambonese from the Moluccas, who seemed to take command, shouted out that the Madurese must be dealt with more harshly. His arm caught a machete and as quick as lightning his shirt turned dark red. The wailing of the Madurese shouting out the greatness of God was unexpectedly frightening. But in the end they were chased away and ran in all possible directions.

On the grass, in the yard, bloodied bodies were strewn about.

A company of Marechaussee, fresh from training in Malang, were brought in to take over from the police, who were considered to have disobeyed orders by firing their rifles, even though only into the air. The police were sworn at and insulted by the Marechaussee and ordered to leave quickly and to pull out the two carriages that had gone into the pond.

A group made up of Madurese, as well as others, charged the compound. It seems they thought the police were still in charge of the operation. Realizing that it was now the Marechaussee they were facing, they hesitated. Some even ran off before entering the compound. Indeed the whole of the Indies feared the Marechaussee, a special command made up of specially chosen troops of the Netherlands Indies Army. They only used rubber truncheons, no firearms or blade weapons. They were famous as a company of fighters.

From the window I saw their leaf-green bamboo hats with the shining copper lion symbol bob up and down amongst the new group of attackers. Their whistles sounded noisily again and again and their truncheons swung round and round, striking and poking, thrusting and thumping. The fight between truncheons and whistles and the other sharp and blunt instruments lasted about half an hour. Two Marechaussee died on the spot.

That time too the protesters were chased away. Darsam was arrested and taken away to who knows where.

After things had calmed down Sergeant Hammerstee banged on the door, wanting to enter. Mama opened it and blocked the way.

"Nyai Ontosoroh?" he asked in Malay.

"I have no business with the Marechaussee."

"This complex is to be guarded by the Marechaussee."

"It's nothing to do with me. No one steps inside my house without my permission."

"I, Marechaussee Sergeant Hammerstee, have come to request permission."

"I do not give permission."

"In that case we will camp in the compound."

Nyai slammed the door shut, locked it from inside, and stood behind it for some time. Looking at me, she said:

"Give in to them once and they'll end up doing as they please. Don't worry. Nothing will happen. They have no papers about this house. They only believe in papers. No matter how tremendous they are, it's all meaningless without papers. Paper determines more, is more powerful." Her voice was bitter.

From the window too I saw Dr. Martinet have his turn in being denied entry by Sergeant Hammerstee after they argued at the main gate for a minute. Their voices couldn't be heard from where I was. His motions indicated that Martinet wished to see his patient, but he was refused. He remained stubborn. But then the doctor climbed aboard his carriage and left.

Now we had to care for Annelies without a doctor.

In the afternoon, Annelies slowly began to awake from her sedation. She opened her big eyes, looked left and right, as if glimpsing the world for the first time ever, then closed them again, and after that opened them again.

"Ann, Annelies," I called.

She looked at me. Her lips opened, pale and bloodless. No voice emerged. I took some chocolate milk and put the glass up to her lips. Silently she drank down almost half a glass, stopped, and sat up in bed. Mama sat silently observing her. Suddenly Mama rose and left the room. At first I guessed she went out back to supervise looking after the cows.

And not long after I heard her voice, half-shouting, in Dutch: "Everybody may go to the Netherlands! Why can't I?"

I took a peep outside in the garden. Nyai was speaking to a Pure-Blood European who stood with hands on hips. His voice was too soft for me to catch his words. The man shook his head for a moment, sometimes shook his finger.

"What does it matter to you if I accompany my own child? I'll use my own money, nobody else's."

The visitor shook his head again.

"Show me the rules where it is written down I can't accompany my own child."

The visitor seemed to move his hands but not his body.

"Smallpox certificate? Health certificate? My child doesn't have one either. On the contrary, she is ill. Get innoculated on board? I can do that too."

I left them both there in the garden. Annelies seemed to be trying to get down out of bed. I helped her to walk. I took her over to the window because that was her favorite place. And we stood there for a long time. But it was impossible to stay silent forever. I forced myself to speak:

"You've never got so far as the mountains, Ann? From the top, you could see all of Wonokromo and Surabaya. We'll go there one day."

You couldn't actually see the mountains. They were covered by clumps of clouds and overcast, like white coffee not properly stirred, made by some lazy person. Low-hanging clouds blocked off the usually black-green forest. Far off in the distance, I couldn't guess how far, lightning flashed, king of the heavens, for a moment. The clouds and grayness then disappeared again to who knows where. Nature was busy with its own affairs.

And beside me, my wife let out a long breath.

Mama entered again. She sat down on the chair, silently, without speaking, as if nothing had happened. When I looked her way, she waved her hand, calling me over. I left Annelies by the window.

"Minke, you must tell her, Minke, her departure is in three days' time."

I had to tell her, because I was her husband. It was indeed my responsibility—a responsibility I still hadn't carried out because of everything that had been keeping me so busy lately. Annelies had to know: We were defeated, crushed, without ever being able to defend ourselves, let alone fight back.

In the distance, nature was still miserable and was becoming more so with every flash of lightning. Under the window our swan pond had been damaged but still hadn't been repaired. A company village, usually visible from the window, and full of playing children, was now still, without any signs of life.

I approached my wife. I put my hands on her shoulders and placed my cheek against her cold cheek. I gathered together all my courage.

"Ann!" She didn't look, neither was there any other response. "Ann, my Annelies, my wife, will you listen to me?"

She ignored me. The fingers of her left hand slowly scratched her neck. That beautiful neck, covered by her crumpled hair, was more perfect than nature outside.

We had only three days left together. She would leave, my darling, my most beautiful doll in the world. What will happen to you later, Ann? And what about myself? Will you be like the lightning outside, flashing for a moment, reigning supreme over all around you, only then to disappear forever? Someone who does not know you at all has suddenly judged you and punished you this way. Someone else again, they too not knowing anything, will separate you from us, and from everything that you love. You were so fragile and pale, Ann. Mama and I had become so thin ourselves.

How sad and moving you were, Ann, so beautiful, but never having had the chance to enjoy your own beauty and youth.

"Don't you want to listen, Ann?" She still paid no heed. "Do you like the mountains over there, Ann?"

There was a hint of a nod, affirming.

"We should have gone riding there, yes, Ann? And Mama will stay at home. We'll go by ourselves, just the two of us, Ann."

Once again she nodded imperceptibly.

"Your favorite horse Bawuk neighs, asking after you all the time, Ann."

She bowed her head. Then, so slowly, she turned around and looked at me and her eyes were like a pair of dreaming daytime star eyes. Her mouth remained mute, and it smelled of medicine.

Mama couldn't hold back her feelings anymore. I could hear her crying and she left the room. About ten minutes later she came in again with another European. He walked straight to where we were standing.

"Government doctor," he said without giving his name, "here to examine the health of Miss Annelies Mellema."

"Mrs.," I retorted.

He paid no heed to me. He led my wife off and sat her down on the bed. He took out a stethoscope from his long coat and began the examination. With eyes popping out, he then checked her blood pressure, tilting his head towards the ceiling. He put the stethoscope back in his pocket. He then examined my wife's eyes. After that he smelled her breath coming from her nose and mouth. He shook his head.

Mama watched all this in silence. The government doctor ordered his patient to lie down.

"Nyai!" he said in the coarsest of Malay. "Why have you allowed this child to be drugged so heavily?"

"Do you wish to leave this house immediately?" replied Nyai in a Malay whose tone was even coarser.

"Don't you understand yet? I'm the government doctor."

"So what do you want?" snapped Mama.

"You can all be charged! Dr. Martinet too! Watch out!"

"Make your charges in your own house, not here. There's no need for you to work your mouth so much here. The door can still swing on its hinges."

The government doctor went scarlet. He turned to me.

"You listen too," he said, using, as with Mama, the coarsely familiar word for *you*. "You can be a witness to this talk, eh?"

"Indeed the door hasn't been nailed shut yet," I said.

Nyai and I went over to Annelies and raised her up so she could eat.

"She's weak, too weak. Let her sleep. Her heart. Don't disturb her," ordered the government doctor.

We got her off the bed and sat her on the settee.

"I'll fetch some food, Ann. Pay no heed to anyone or anything."

She nodded weakly.

The doctor approached me in a threatening manner, and indeed, threatened me:

"You're trying to oppose my orders, *hey!*"

"I know my wife better than any outsider," I answered in Malay, without looking at him.

"Good," he said and left the room. "Watch out!"

"Why won't you speak, Ann?" She was still silent. "Will you listen to me, Ann? That puffed-up doctor has gone. Don't be afraid."

I followed her eyes, which were directed out the window, and let my eyes head for the mountains, which were still covered by clouds. Mama observed my actions without speaking.

Annelies chewed slowly, very slowly, each time hesitating at swallowing.

From behind me I heard Mama speak, more to herself:

"Maurits before dug up past blood sins. Now he demands the wages of those sins. Before I thought he was some holy prophet. . . ."

"There's no use in remembering, Ma," I said without turning to look at her.

"Yes, memories sometimes torture. Indeed there's no use in remembering. Have you told her, child, Nyo?"

"Not yet, Mama."

"Speak Ann. You've been silent so long."

Annelies looked at me. She smiled. Smiled! Annelies smiled! Mama opened her eyes wide in amazement. You're getting better, Ann, I exclaimed in my heart.

Mama rose from her place, embraced her child, kissed her, mumbling:

"The sadness disappears because of your smile, Ann, for your husband too. It's been too much, you not speaking all this time." And her tears streamed forth.

Annelies blinked slowly, so slowly, as if she didn't really want to open her eyes again.

Dr. Martinet had once said: Her difficulty was that she wanted to consolidate tightly what was already there. She didn't want to let go of what she had already seized hold of. But a crisis could one day occur which would force her to let go of everything and she would no longer care about anything that happened to her. Was this the stage my wife was at now? I didn't know. Dr. Martinet was not allowed to see her. His last words had been: If Annelies could be convinced to surrender to the situation, she would be safe. And how were things now? I didn't know. Mama didn't know. How far away you were, Doctor!

While she was under Martinet's care she was, so he explained, still clinging tightly to things as they were. We'd all been defeated, he said, all our efforts had failed, while Annelies didn't want to understand any of this. She did not seem to rebel, but within her there was disarray as an uncertain war raged. Only being sedated could save her from psychological damage. If not, it might happen that nothing again would have value for her. Or, on the other hand, she could become worthless to anyone. Remember Mr. Mellema. So, if she becomes conscious, talk to her without pause, about things beautiful, good; things of hope and pleasant things.

And now it was my job as husband to tell her the bitter truth: Three more days! And she wouldn't be drugged. Dr. Martinet was not allowed to see her.

The doctor had also once said: Annelies has passed the crisis period. He said that just a little while before we married. Now a new crisis was upon her. This time also, he said, I'm not her doctor, but you, her husband, the person she loves. You must try to leave with her for the Netherlands. Nyai will be able to pay for the trip: one hundred and twenty guilders. That wouldn't be too expensive for her.

They had forbidden us to escort her.

Try, said Dr. Martinet, by whatever means. Don't let the life of your wife be wasted. She will not be able to live without you. You now are the only thing to which she clings.

I felt that I had done everything within my power, and I had been defeated. The Amsterdam District Court could not be opposed. The Surabaya court for Europeans had pronounced that Mama and I had no connection with my wife. Nyai herself cleverly ordered Panji Darman, formerly Jan Dapperste, to set sail "to look after the spice business" in the Netherlands and to befriend Annelies as my representative. Nyai had forbidden him to come to Wonokromo to avoid suspicion falling upon him. And a Netherlands Transport Company agent brilliantly placed him in a second-class cabin next door to Annelies's cabin. The agent was the one who also got his health certificate predated.

The face of my wife was like chiseled marble, as if the nerves

351

in her face had been severed from her brain. There was no move-ment, no expression whatsoever, and she still didn't speak. I had tried from every angle, used every approach to tell her of the day of her departure. It had all failed.

She ate no more than four spoonfuls, then did not want to open her mouth again. I don't know how many times Nyai ner-vously walked in and out of the room. Once when the room was empty I embraced my wife and I forced myself to have the courage to whisper in her ear:

"Ann, we're defeated, Ann; we were to go with you on the boat to the Netherlands, but they will not allow us. Ann, do you hear me, Ann?"

She still did not respond.

"I don't know what you're thinking. But you need to know, Ann: Jan Dapperste will be there in place of Mama and me. In three days he will escort you as you set sail for Europe. Don't be afraid, Ann. Once you've arrived there, Mama and I will follow."

And Annelies still paid no attention. Yet I had carried out my duty as her husband, a duty by no means fully accomplished: She still hadn't responded. How many times would I have to repeat the information? I kissed her. Still no response. Perhaps Dr. Mar-tinet was right? She had passed the crisis point and was now beginning to set everything free?

For the umpteenth time Mama entered. This time she handed over a telegram from Herbert de la Croix and a letter from Mother.

The assistant resident of B—— passed on his regrets that the attorney he had sent had failed. He shared our sadness and ex-pressed his sympathy for us. In his rather long telegram he also stated that the Amsterdam court's decision was unjust. He had telegraphed the government-general saying that he would resign from his position if the court's decision was carried out. He had also sent a telegram to protest to the Ministry of Justice, all to no avail—they didn't even bother to answer. So he was going to resign and return to Europe with Miriam.

And Annelies herself? She had still lost her interest in every-thing. And I talked and talked, told story after story. And she still wouldn't talk. Perhaps she wasn't even listening. I carried her back to bed and I laid her down, and myself lay down beside her. It was lucky I knew so many stories as well as the legends of my ances-

tors. And yet I had already told them all. The European story of
Prince Genevieve at least four times, *Gulliver's Travels* twice,
Baron von Munchhausen twice, and Little Duimpje perhaps more
than four times. And there were still the mouse-deer stories. My
voice was already hoarse. I still added to all these stories, others of
my own experiences that were a bit funny.

Embracing my wife, I told fairy tale after fairy tale. I brought
my mouth close to her ear—something she liked.

When I awoke, the night had passed, the room was bright
with the rays of the sun. Yet my tiredness hadn't been chased
away by the sleep whose length I didn't know. And all of a sudden
I realized: Annelies was embracing me, kissing me, and caressing
my hair. I sat up in a hurry.

"Ann, Annelies!" I shouted. I held her wrist and I felt how
her pulse was not as slow as the day before.

"Mas!" she answered.

Was it true that my Annelies was beginning to speak? Or was
I just dreaming? I rubbed my eyes. Dream, don't bother me like
this! But my eyes saw my wife smiling. Her face was pale, her
teeth dirty. And her eyes did not share the smile.

"Ah, Annelies, my Annelies! You're well again, Ann!" I embraced her and kissed her. My efforts and labor all these last days
hadn't been wasted.

"Food is ready, Mas, let's eat," she said gently, exactly as she
used to.

I looked at her. Was Dr. Martinet right: Had she been flung
off balance, shocked, so her mind was no longer able to work
properly? I looked closely at her eyes. And those eyes were sad.
Her lips still smiled but her eyes did not; instead, it was as if she
had gone cross-eyed.

"Mama!" I shouted. "Annelies is well again."

And Mama didn't appear.

And without having washed first, I sat in that room facing the
dinner.

There was no spoon or fork or plate before me. Only before
Annelies. Had she lost her mind or was I to eat alone?

She began to spoon up the food and feed it to me.

"I can eat myself, Ann. It's you who must eat, let me feed
you."

She didn't eat, but fed me again. And I had to chew and

swallow. I must not offend her—I knew that with certainty—and so I ate until I was full.

"Why are you feeding me like this?"

"Once in my life let me feed my husband." She went silent and did not want to speak again.

20

This day—the last day.

The business had come to a total standstill. The Marechaussee were stopping everybody from entering our front grounds. All we were allowed to do was milk the cows.

Mama's protests were ignored.

"It's not costing Nyai anything," they retorted. "It's the people of the Netherlands who are suffering the losses."

Many letters arrived. There was no opportunity to reply. There was no time really to read them either. The papers sent by Nijman piled up untouched.

Mama, I, and especially Annelies, were not allowed to leave the house, except to wash and to go to the toilet. So we were under house arrest.

The Marechaussee soldiers only left their tents in the compound to chase off people who gathered on the edge of the road who were expressing their sympathy for us, perhaps, or were only there to have a look.

Annelies looked more like her usual self, although she was skinny and pale and her eyes were dead.

"Tell me about Holland according to Multatuli's stories," she suddenly asked.

"There was a country on the edge of the North Sea. . . ." I began as best I could. "Its land was low-lying, so it was called the Land of Low Country—Netherlands, or Holland." When I reached that point I could find no way to continue. Those dreaming eyes of hers, still sad, looked at me so strangely, as if I were some new kind of blue-tailed lizard that she was seeing for the first time in her life. "Because the land was so low-lying, people became bored with repairing their dikes, so it became their habit to leave their country, to wander, Ann, to admire those other countries with their mountains. Then to conquer them, of course. In those high countries they made the people low. Nobody was allowed to even approach the height of a Dutchman."

"Tell me about the sea."

A European woman in white clothes and hat entered without knocking. Nyai and I let her do so; during the last few days anyone and everyone had been coming in and out of our rooms. She would only annoy us anyway.

"In four more hours you will be sailing across the sea, and more sea, and more sea, dear," the new arrival said, taking over my job. "There are more fish than you would ever imagine. Waves, ripples, swell, spray, and foam. Miss will be sailing on a big ship, beautiful, crossing the ocean, dear, entering the Suez canal, passing by other ships on the way. When they pass, dear, the ship's whistle will blow. The others will blow theirs as well. Have you seen Gibraltar? Ah, you will pass by that town of coral too. And after that, a few days later, you will set foot on the land of your ancestors. Its sands are a shining, golden yellow, flowers everywhere, all as miss wishes. It makes people happy. Soon autumn will arrive. Leaves will fall. . . . How happy you will be, looked after by your own brother—a scholar, engineer, well-known, honored, and respected by people. How happy you will be . . . if you don't like it, perhaps in one or two years you'll be able to decide your own life. Yes, miss, just one or two years . . ."

"Mas, I like the waves, and foam, and swell more than ships and the Netherlands . . ."

"No, dear," the newcomer cut in, "in the Netherlands you will find everything. Everything that miss wants she will be able to find there."

356

"Mas, is there anything lacking here?"

"No, Ann. You have everything here. You are happy here."

"If the Netherlands has everything," Mama added angrily, "why have Europeans come out here?"

"Don't make my work more difficult, Nyai. Get her clothes ready."

"No, not just her clothes"—Mama began to become irritable—"her jewelry, also her bank book, also the letter of acknowledgment of her father, and the prayers of her mother and husband."

"Mama," Annelies cut in, "does Mama remember Mama's story before. . . . ?"

"Yes, Ann, what story do you mean?"

"Mama left home forever. . . ."

"Yes, Ann, why?"

"Mama took an old brown tin suitcase."

"Yes, Ann."

"Where is it now, Mama?"

"Stored in the attic, Ann."

"I want to see it."

Mama went to fetch it.

"It's nearly time, miss," the European woman cut in.

Neither Annelies nor I responded. And Mama brought a small, brown, rusted, dented suitcase. Annelies took it quickly.

"With this suitcase, I will go, Mama, my Mama."

"It's too small and horrible. It's not fitting, Ann."

"Mama, it was with this suitcase that Mama left that time resolving never to return. This suitcase weighs too heavily on Mama's memory. Let me take it, Mama, along with the burdensome memories it contains. I will take nothing but the batik kains made by Minke's mother. Only this suitcase, Mama's memories and, Mother's batiks, my wedding clothes, Mama. Put them in. My devoted obeisances to Minke's Mother. I will go Mama. Don't remind yourself of all those things from the past. That which has passed, let it pass away, my Mama, my darling Mama."

"What do you mean, Ann?"

"Like Mama before, Mama, I too will never return home."

"Ann, Annelies, my darling child," cried Mama and she embraced my wife. "It's not that Mama didn't try, Ann, it's not that I didn't defend you, child. . . ."

Mama sank into remorseful sobbing. So did I.

"We both did all we could, Ann," I added.

"Don't, don't cry, Mama, Mas; I still have a request, Mama, don't cry."

"Tell us, Ann, tell us." Mama began to wail.

"Mama, give me a little sister, Mama, a little sister, who will always be sweet to you . . ."

Mama wailed even more.

" . . . so sweet, Mama, not causing you trouble like this daughter of yours . . . until . . ."

"Until what, Ann?"

" . . . until Mama no longer misses Annelies."

"Ann, Ann, my child, how can you talk so. Forgive us that we could not defend you, forgive us, forgive, forgive."

"Mas, we were happy together?"

"Of course, Ann."

"Remember only that happiness, Mas, nothing else."

"Come on!" shouted out an Indo from the door. "We're two minutes late already."

"Come, dear, miss," the European woman guided Annelies.

Annelies at once submerged into muteness and disinterest. Her momentary dignity suddenly disappeared. She walked slowly out of the room and down the stairs, under the guidance of the European. Her body seemed so broken and exhausted.

Mama and I ran up to support her, to take over from the woman. But the Indo man and the European woman stopped us.

Marechaussee had congregated at the bottom of the stairs.

And we were kept away! So we were only able to watch our beloved Annelies led away like a cow, step by step.

Perhaps this was how Mama's mother felt when she was treated that way by Mama because she was unable to defend Mama from Mellema. But how did Annelies feel? Was it true she had let go of everything, her own feelings too?

I no longer knew anything. Suddenly I heard the sound of my own crying. Mother, your son had been defeated. Your beloved son did not run, Mother; he is no criminal, even though he's proven incapable of defending his own wife, your daughter-in-law. Is this how weak a Native is in the face of Europeans? Europe, you, my teacher, is this the manner of your deeds? So that even my wife, who knows so little about you, lost all belief in her little world—a world incapable of providing security even for her. Just one person.

I called after her. Annelies didn't answer, nor look back.

"I'll be following soon," I shouted.

No answer, no look back.

"I too, Ann; have courage, Ann!" called out Mama, her voice hoarse, almost unable to escape from her throat.

Again no answer, no look back.

The gate was opened. A government carriage was waiting, hemmed in on either side by mounted Marechaussee. Mama and I weren't allowed past the door.

For a moment we could still see Annelies being helped into the carriage. She still didn't look this way, didn't make a sound.

The door was closed from outside.

The sound of the carriage wheels grinding over the gravel could be faintly heard fading away into the distance, finally disappearing. Annelies was setting sail for where Queen Wilhelmina sat on the throne. Behind the door, we bowed our heads.

"We've been defeated, Mama," I whispered.

"We fought back, child, Nyo, as well and honorably as possible."

Buru Island Prison Camp
Spoken, 1973
Written, 1975

AFTERWORD

During the first six or seven years on Buru Island, prisoners were not allowed reading material, except for a few religious texts. To be found with illicitly obtained material could result in severe punishment. One prisoner, while out working in the fields, came across a piece of torn newspaper. It had been used to wrap nails in and was full of holes. When the prison guards found the newspaper on him, they took him away to the cell block. Three days later his body was found floating in a nearby river, his hands bound behind his back.

Pramoedya Ananta Toer's commitment to literature has clearly survived this hostility to the printed word. Of Javanese descent, he has been writing since the armed struggle for Indonesian independence, which broke out in 1945. Many of his early works were written while languishing in a Dutch prison. He was captured by the Dutch Colonial Army during its attack on Jakarta in 1947, two years after the Revolutionary War started, and was released in 1949. After Indonesia's independence, Pramoedya began an active life in the literary world, producing several novels and many short stories. In the late fifties, he began his serious study of history and lectured in history and journalism at universities and academies in Jakarta. In the sixties, he entered the fierce polemics on the role of literature in society and attacked works that he felt ignored social problems and the political crisis the country was facing. He was a leading member of the People's Cultural Institute, many of whose members were close to the Indonesian Communist party. Pramoedya himself was not a member and did not write in the party's daily newspaper, but rather became the cultural editor for another independent newspaper, *Bintang Timur* (the *Eastern Star*), which reprinted many early writings of the pioneers of Indonesian fiction and journalism.

Pramoedya's interest in history continued to flourish and the *Eastern Star* published many of his articles about the period of Indonesia's national awakening. A part of this historical work included carrying out the research and contemplating the framework for a series of historical novels about the birth of national consciousness in Indonesia. The period to be covered was the turn of the century, 1890 to 1920. With the help of many of his students, a great deal of work was done and material collected.

Then, in September 1965, a coup attempt took place in Indonesia. Disorder and confusion occurred in Jakarta and throughout the country. The coup attempt was blamed on the Indonesian Communist party. The Indonesian army took control to restore order, but in the process almost all center, left-wing, and progressive political groups suffered persecution. It has been estimated that as many as 500,000 or more people were killed in this nationwide purge. Along with thousands of others, Pramoedya was arrested (though even after fourteen years in jail, he was never tried). His library, including all his notes and the results of his research, was burned on the spot. In 1960 the Sukarno government banned his history of the role of the Chinese in Indonesia; in 1966, with the rise of Indonesian McCarthyism, a blanket ban was placed on all his earlier books—a ban that never has been revoked.

It was only eight years later that Pramoedya was able to begin to put down on paper what had been lost in 1965. He had already, each day before roll call, presented an oral version of the novels— to ensure, he says, that the story would not be lost if the chance to produce a manuscript never arrived. In 1975, after two years of writing from memory, he finished his series of novels.

This Earth of Mankind was published in Jakarta as *Bumi Manu-*

sia a year after Pramoedya's release from Buru Island concentration camp in 1979. Soon after, its sequel *Anak Semua Bangsa* (*Child of All Nations*) was published. Both novels became best sellers in Indonesia, as reviewers hailed Pramoedya's return to the nation's literary life. However, in May 1981, both books were banned in Indonesia. The government accused the books of surreptitiously spreading Marxism-Leninism—surreptitious because, they claimed, the author's great literary dexterity made it impossible to identify actual examples of this Marxism-Leninism. Both the third and fourth volumes of the tetralogy have also now been banned in Indonesia.

In its letter to the prosecutor general calling for a revocation of the ban on *This Earth of Mankind,* the Jakarta Legal Aid Institute reminded the authorities that it was no longer rational to try to isolate Indonesia from so-called foreign ideologies. This was because free exchanges of ideas came automatically with the social intercourse between nations, the international coming and going of reading materials, the progress in communications technology, and the coming together of persons in the world's cultural centers. Such exchanges keep the world dynamic, generating pressures for change and demands for debate everywhere. They did so to no less an extent one hundred years ago in the colonies of all the European imperial powers.

Max Lane
1991

GLOSSARY

assistant resident	For each regency there was a Dutch assistant resident in whose hands power over local affairs ultimately resided.
Babah	A term of address referring to Chinese shopowners, this also has connotations of "boss."
Betawi	The Malay name of Batavia, the capital of the Dutch East Indies, now Jakarta, the capital of Indonesia
blangkon	A traditional Javanese headdress made from batik and worn mainly by the nobility or those with pretensions to an elite status
bupati	This is the title of the Native Javanese official appointed by the Dutch to administer a region. Most bupatis could lay some claim to noble blood.
destar	An East Javanese headdress; a kind of headband

Gus	A term of affection used by parents toward their male children among the families of the Javanese aristocratic elite
kabaya	A Javanese woman's traditional blouse, used always in combination with a sarong
kain	This traditional dress worn by Javanese women is a kind of sarong wrapped tightly around the waist and legs.
keris	A traditional Javanese dagger
Marechausee	The elite troops of the colonial army in the Netherlands Indies
Mas	This Javanese term of address literally means "older brother." Used by a young woman toward a man, it indicates an especially close, respectful affection. It can also be used between men to indicate respectful friendship, by a sister to her older brother, and also by a wife to her husband.
Ndoro	An honorific used by a lower-class person when speaking to someone in the feudal class or one of similar status
Noni, Non	Miss
Nyai	The Native concubine of a Dutch or European man in the Indies
Nyo	The abbreviated form of sinyo, used by the Javanese to address young Dutch boys
patih	The chief executive assistant and secretary of a bupati
peci	A small black velvet cap, originally a sign of Islam
priyayi	A member of the Javanese aristocracy, many of whom became the salaried administrators of the Dutch
Raden Ayu	The title for aristocratic Javanese women, especially the first wife of a bupati

Raden Mas — Raden and mas are titles held by the mass of the middle-ranking members of the Javanese aristocracy; raden mas is the highest.

regency — Generally made up of more than five districts, a regency was an area under native administration. The head of the regency was the bupati. His supervisor was a Dutch assistant resident. Above the assistant resident was the resident, who had responsibility for many regencies.

sanggul — This bun at the back of the head is a traditional Javanese hairstyle.

sarong — A wrap garment worn by both men and women

Sinkeh — A term used to refer to a new Chinese immigrant

Sinyo — The Javanese form of address for young Dutch and Eurasian men or Europeanized Native young men, from the Portuguese word *senhor*

tayub — This is a folk dance in which the male partner is normally chosen from among the audience by the professional female dancer. Most tayub dancers were from the lower social strata.

tuan — A Malay word meaning master or sir

Tuan Besar Kuasa — Great powerful master, a term used for a Dutch administrator or other powerful official